Utilization of Phytol in Forages for Ruminant Production

叶绿醇在反刍动物饲养中的应用

Renlong Lyu

吕仁龙 / 主编

中国原子能出版社

图书在版编目（CIP）数据

叶绿醇在反刍动物饲养中的应用 / 吕仁龙主编. -- 北京：中国原子能出版社，2022.5
ISBN 978-7-5221-1947-2

Ⅰ.①叶⋯ Ⅱ.①吕⋯ Ⅲ.①植醇—应用—反刍动物—饲养管理 Ⅳ.①S823

中国版本图书馆 CIP 数据核字（2022）第 074141 号

内容简介

叶绿醇是叶绿素分子支链，在反刍动物瘤胃内通过生物氢化作用后以游离态的存在，其再经过瘤胃微生物作用会生成植烷酸。这些植烷酸会随着代谢分布在瘤胃、血液中，参与代谢沉积到奶和肉中。植烷酸具有调节胰岛素，促进脂肪酸氧化的作用，对动物机体及畜产品对人体健康都具有积极影响。青绿饲料是反刍动物重要饲料来源，优质的青绿饲料不仅蛋白质，纤维等营养成分丰富且功能性物质，如植物光合色素含量也较为丰富。本专著特别着眼于植物光台色素中色素中的叶绿素，阐述了叶绿素含量在青绿饲料中的变化，分解及其对反刍动物的影响。

叶绿醇在反刍动物饲养中的应用

出版发行	中国原子能出版社（北京市海淀区阜成路 43 号 100048）
责任编辑	张　琳
责任校对	冯莲凤
印　　刷	北京亚吉飞数码科技有限公司
经　　销	全国新华书店
开　　本	710 mm × 1000 mm　1/16
印　　张	8.75
字　　数	166 千字
版　　次	2023 年 3 月第 1 版　2023 年 3 月第 1 次印刷
书　　号	ISBN 978-7-5221-1947-2　　定　价　98.00 元

网　　址　http://www.aep.com.cn　　E-mail:atomep123@126.com
发行电话：010-68452845　　　　　　版权所有　侵权必究

Editor in Chief:
Renlong Lyu
(Tropical Crops Genetic Resources Institute,
Chinese Academy of Tropical Agricultural Sciences, China)

Associate Editor:
Taketo Obitsu
(Graduate School of Integrated Sciences for Life,
Hiroshima University, Japan)

Editorial Staff:
Mabrouk Elsabagh
(Faculty of Agricultural Sciences and Technologies,
Niğde Ömer Halisdemir University, Turkey)

Hanlin Zhou
(Zhanjiang Experimental Station of Chinese Academy of Tropical
Agricultural Sciences, China)

Peng Wang
(College of Animal science, Jilin University, China)

前言 PREFACE

  天然光合色素广泛存在于自然界,叶绿素、叶黄素、β-胡萝卜素不仅让动植物生命绚丽多彩,更与人类活动、机体健康息息相关。光合色素具有极强抗氧化特性,对生命体具有积极影响。本书主要着眼叶绿素资源,以牧草和反刍动物为模型,解析叶绿素在饲用牧草不同加工条件下的变化规律,探究叶绿素在反刍动物体内的代谢机理。

  一个叶绿素分子的核心部分是卟啉环(Porphyrin ring),其功能是光吸收;其他部分是一条很长的脂肪烃侧链,称为叶绿醇(Phytol),叶绿醇在反刍动物体内经过生物氢化和酶的共同作用会产生具有促进脂肪酸氧化和协同胰岛素作用的植烷酸(Phytanic acid)。植烷酸对人类机体健康具有积极影响,由于其不能在哺乳动物体内合成,因此摄取反刍动物产品(肉、奶)是人体获得植烷酸的唯一来源。探索叶绿素和植烷酸在动物体内变化规律和转化机理对生产高品质(高植烷酸)畜产品具有重大意义。

  深入挖掘并充分利用天然活性成分来提升反刍动物产品附加值是未来研究的一个重要方向,有助于进一步理解动植物生命活动的关联,并揭示微观生命规律。

  近年来,功能性食品受到广大消费者青睐。消费者的关注点逐渐从感官向微观转变,在选择终端商品时,更重视营养成分、微量元素以及活性物质含量等指标。随着研究深入和市场商品化不断进步,功能性食品的标准也在不断变化。希望通过本书的出版,能够在未来引起人们对于如何定义和生产功能性畜产品标准的思考。

本书在写作整理过程中一定存有不足之处,希望广大读者及同行领域专家给予指导和建议。

作　者
2022 年 5 月

目录 / contents

Introduction of phytol ··· 1
  1. Physical-chemical property of phytol ······························· 1
  2. Effects of phytol on mice ·· 2
  3. Effects of phytol on finishing pigs ···································· 5
  4. Effects of phytol and its metabolites ································ 6

Chapter 1   General introduction ··· 13
  1.1 General background ··· 13
  1.2 Photosynthetic pigments and phytol in herbages ············· 14
  1.3 Decomposition pathways of chlorophyll ························· 15
  1.4 Functional compounds in ruminant products··················· 15
  1.5 Objectives ·· 17

Chapter 2   Changes of photosynthetic pigments and phytol in herbages ·· 19
  2.1 Changes of photosynthetic pigments and phytol content at different fertilization levels in fresh herbage and hay ······ 19
  2.2 Changes in carotenoid, chlorophyll and phytol contents in Italian ryegrass during ensiling ···································· 29
  2.3 Effects of nitrogen fertilization and harvesting stage on photosynthetic pigments and phytol contents of Italian ryegrass silage ··· 37
  2.4 Effects of lactic acid bacteria addition on photosynthetic pigments of ensiled Italian ryegrass grown under different nitrogen fertilization levels ·· 52

2.5 Changes of chlorophyll and phytol contents in different harvesting height King grass before and after ensiling ··· 60

Chapter 3 Phytanic aicd production in the rumen ············ 73

3.1 Effects of fertilization levels and harvesting stages of fresh herbages on ruminal phytanic acid production in vitro ··· 73

3.2 Effects of fertilization levels and harvesting stages of Italian ryegrass silages on ruminal phytanic acid production in vitro ··· 81

Chapter 4 Effect of forages sources on phytanic acid content in milk of dairy cows ············ 89

4.1 Introduction ············ 89
4.2 Materials and methods ············ 90
4.3 Results ············ 92
4.4 Discussion ············ 92

Chapter 5 General discussion ············ 98

5.1 Factors for the changes of photosynthetic pigments and phytol in herbages ············ 99
5.2 Factors for the changes of phytanic acid in the rumen ··· 100
5.3 Factors on the changes of phytanic acid content in milk ··· 101
5.4 Conclusion ············ 102

Chapter6 General summary ············ 105

References ············ 109

Abbreviations ············ 127

Introduction of phytol

1. Physical-chemical property of phytol

Phytol is a chain diterpenoid oxygen-containing compound, which is an unsaturated primary alcohol. It is widely distributed in plants and is a moiety of chlorophyll, which could be obtained by hydrolysis of chlorophyll. Phytol has E and Z isomers, both of which exist naturally. E-phytol can be extracted from silkworm sand, which is a colorless transparent oily liquid with a boiling point of 202 ~ 204 ℃ (10 mmHg). It is almost insoluble in water and soluble in ethanol, acetone and other organic solvents.

Phytol is also an isoprenoid alcohol bound in ester linkage to chlorophyll, which is the most abundant photosynthetic pigment in plants. During leaf aging period, large amounts of phytol are released by chlorophyll degradation. However, the pathway of phytol decomposition in plants is unclear. It was hypothesized that phytol degradation in plants might involve its oxidation into the long-chain aldehyde phytenal. The aldehyde quantification after derivatization with methylhydroxylamine was determined by C-MS, in addition phytenal was found in leaves, while other long-chain aldehydes (phytanal and pristanal) were barely detectable. Analysis of phytochemicals accumulated during chlorophyll degradation in PA1 mutant showed that phytochemicals were phytochemicals produced during chlorophyll degradation. The increase in plant-based fatty acids was even stronger in mutants of other leaf alcohol metabolites including vte5-2 (tocopherol deficiency) and pes1 pes2 (fatty acid leaf alcohol ester deficiency).

2. Effects of phytol on mice

2.1 Effects of phytol on average body weight of mice

At present, there are a few reports about the effects of phytol on animal growth and development, and the research results are different. Mackie et al. (2009) found that the weight of mice fed with diets containing phytol (0.5%, 0.1%) for 3 weeks decreased significantly. Hashimoto et al. (2006) also found that adding 0.5% phytol in the diet could reduce the feed intake mice and slow down the weight gain of mice. The results of Lin et al. (2018) study showed that compared with the control group, the average body weight of mice fed with 0.50% phytol for 2 weeks could be significantly reduced. These experimental results are basically consistent with previous results, and their mechanism needs further research.

2.2 Effects of phytol on metabolic enzyme activity and triglyceride content of gastrocnemius muscle in mice

Muscle metabolism can be roughly divided into oxidative, glycolic and intermediate types. The metabolism of type I muscle fiber is mainly oxidative type, and type II muscle fiber is mainly glycolysis type (Wang Li, 2016). Under the same conditions, the metabolic type of muscle is related to its own status, and the activities of lactate dehydrogenase (LDH), succinate dehydrogenase (SDH), hexokinase (HK) in muscle can reflect the metabolic status of muscle. SDH is one of the junctions connecting oxidative phosphorylation and electron transport, and could provide electrons for the aerobic and productive respiratory chain of eukaryotic mitochondria and a variety of prokaryotic cells. Its activity can generally be used as an indicator to evaluate the running degree of the tricarboxylic acid cycle. SDH, different from other enzymes in the tricarboxylic acid cycle, is the only enzyme embedded in the mitochondrial intima, which is an important part of the mitochondrial intima (Zhang et al., 2007). Some studies showed that those athletes trained in speed and explosive force have selective hypertrophy of fast twitch fiber, while athletes trained in endurance have selective hypertrophy of slow switch fiber which contained more slow muscle and the activity of SDH increased significantly (Saltin et al., 1977). After measured the activity of SDH in various muscle fiber types of mice diaphragm. It was reported in the study of Sieck et al. (1995) that the activity of SDH in type I and type II a muscle fiber was the highest. And it was also found in the study of He et al. (2001) that SDH activity of three muscle fibers in human skeletal muscle, type I had the highest activity, followed by type II a and type II b. HK is the first enzyme in glycolysis, which converted glucose to glucose-6-phosphate. Glucokinase is an isomer of HK. LDH is a binding protein whose activity can reflect the activity of anaerobic glycolysis in cells (Liu, 2001). Some studies have shown that comparing the activities of HK and LDH in red femoris muscle (II a more), white femoris muscle (IIb more) and soleus muscle (I) of guinea pigs, the LDH activity in soleus muscle was the lowest and the LDH activity in white femoris muscle was the highest, while the HK activity showed opposite tendency (Gillespie et al., 1970; Peter et al., 1971). After comparing the activities of LDH and HK in white hemimembrane muscle and soleus muscle of guinea pig, the same result were shown in the study of Peter et al. (1972). These results suggested that the activities of SDH and HK in I muscle fiber are higher than those in II muscle fiber, while the activities of LDH are lower than those in II muscle fiber. And the results of Lin et al. (2018) study showed that the addition of phytol could improve the activities of HK and SDH in mice gastrocnemius muscle, which indicated that phytol could improve the activity of oxidative metabolic enzymes in muscle, thus improve the oxidative metabolism level of muscle. The results of Lin et al. (2018) study showed that TG content of gastrocnemius muscle in the treatment of adding 0.50% phytol was significantly lower than that in control group. Although the study of Serra et al. (1998) showed that intramuscular fat content was higher in muscles dominated by oxidative metabolism, this result needed be further explored. At present, the studies of Seideman et al. (1986) has proved that there was no correlation between intramuscular fat and muscle metabolism type. The lipid content of muscle fibers in oxidized muscle was higher than that in glycolysis muscle, while the lipid content in muscle fiber-was lower than that in adipocytes, therefore, the content of intramuscular fat was mainly determined by muscle adipocytes. Meanwhile, it was found that the TG content in liver of mice decreased by 3 times and 4 times after fed the mice with the diet containing 0.2% and 0.5% phytol for 4 weeks respectively (Hellgren, 2010), which were similar to previous studies that phytol had the effect of reducing the TG content in tissues.

2.3 Effects of phytol on muscle fiber types of mice gastrocnemius muscle

Muscle fiber is the basic unit of muscle, and the composition difference of muscle fiber type is an important factor of meat quality difference (Karlsson, 1999), which has an important relationship with meat quality indicators such as tenderness, pH, color difference and water-holding capacity, etc. (Jeong et al., 2012; Kim et al., 2013; Lee et al., 2010; Choe et al., 2008; Ryu et al., 2008). The transformation among different muscle fiber types might be the key to meat quality improvement. High proportion of type I muscle fiber showed higher pH24h and oxidase activity, lower enzyme activity and drip loss (Gil et al., 2003). Some studies have shown that increasing the proportion of type II muscle fiber in pigs could increase pH decline, meat color brightness, meat greyish-white, cooking loss, protein denaturation degree and decrease of water-holding capacity after slaughter. (Choi et al., 2007; Ryu et al., 2005; Bowke et al., 2004; Sazili et al., 2005). The results of Lin Xiaqing et al. (2018) showed that adding 0.5% phytol could increase the proportion of type I muscle fiber, which suggested that phytol may improve meat quality.

2.4 Conclusion

In summary, adding phytol in diets could improve the relative weight, the activities of HK and SDH and the proportion of type I muscle fiber in gastrocnemius muscle of mice.

3. Effects of phytol on finishing pigs

The results of Wang Jianjun et al. (2014) showed that the diets containing different ratios of phytol had no significant effects on performance and carcass traits of finishing pigs ($P>0.05$), however, the diet containing 0.025% of phytol could significantly increase the longissimus dorsi pH24h ($P<0.05$). The addition of 0.05% and 0.1% phytol significantly increased marbling score of longissimus dorsi of finishing pigs ($P<0.05$), however, the addition of phytol had no significant effects on pH45min, shear force and meat ($P>0.05$). In addition, the diets containing 0.025% and 0.05% phytol significantly increased the activity of succinate dehydrogenase and decreased the activity of lactate dehydrogenase in longissimus dorsi ($P<0.05$), however, there was no significant effect on the activity of hexokinase.

The addition of 0.025% and 0.05% phytol significantly increased the proportion of type I muscle fiber in longissimus dorsi ($P<0.05$), however, there was no significant effect on the proportion of type IIa and type IIb muscle fiber ($P>0.05$). In conclusion, above results indicate that the diets containing phytol could increase eye muscle area and marbling and promote muscle fiber type transformation, which provided related basis for further studies on developing new feed additives to improve meat quality.

4. Effects of phytol and its metabolites

4.1 Regulation of phytol and its metabolites on adipocyte differentiation and glycolipid metabolism

4.1.1 Regulation of differentiation of white adipocytes

The study of Schluter et al. (2002) found that phytanic acid could successfully induce the differentiation of 3T3-L1 cells and human pre-adipocytes into white adipocytes. Under the condition of culture medium without differentiation induction, 70% of 3T3-L1 pre-adipocytes could be induced to differentiate after treated with 40 μmol/L phytanic acid for 2 weeks, and more than 85% of 3T3-L1 pre-adipocytes could be induced to differentiate after treated with 80 μmol/L phytanic acid for 2 weeks. The results of gene expression showed that the expression level of aP2 mRNA in cells treated with 80 μmol/L phytanic acid was almost the same as that in cells treated with the optimum inducer. After comparing the human white pre-adipocytes inducer adding 1 μmol/L rosiglitazone (BRL49653, peroxisome Proliferative activator receptor γ (PPAR γ) agonist) with that adding 80 μmol/L phytanic acids, it was found that the adipocyte differentiation polyester could reach 65% of BRL49653. However, Heim et al. (2002) showed that the effect of 50 μmol/L phytanic acid on C3H10T1/2 differentiated polyesters of mice embryonic fibroblasts was limited, which suggested that the regulation of phytanic acid on different cell differentiation polyesters had cell selectivity.

4.1.2 Regulation of differentiation of brown adipocytes

Phytol and its metabolites can induce the differentiation of primary brown adipocytes into mature adipocytes. It was found that less than 1 μmol/L phytanic acid could affect the differentiation of brown adipocytes, existed 25% of cell differentiation polyester and the expression of aP2 mRNA increased 3.1 times. In addition, the study of Schluter et al. (2002) found that phytanic acid was an effective activator of uncoupling protein 1 (UCP1), and catalase activity increased by about 3 times after rat UCP1 and catalase reporter plasmid HIB-1B cells were transfected with 20 μmol/L phytanic acid. Crisan M. et al. (2008) reported that UCP1 mainly exists in brown adipose tissue and is an important protein for non-shivering thermogenesis in rodents, as well as an important factor affecting animal energy balance. The study of Seedorf U. et al. (1998) explained that increasing phytanic acid in mice could increase feed intake, while there was no effect on their body weight. These results suggested that phytol and its metabolites might regulate energy balance by activating UCP1 protein.

4.1.3 Regulation of glucose and lipid metabolism in liver

The study of Heim et al. (2002) showed that after the primary rat hepatocytes treated with 100 μmol/L palmitic acid docosahexaenoic acid (DHA) and phytanic acid for 24 h, palmitic acid and DHA inhibited glucose uptake by liver cells, while phytanic acid significantly increased glucose uptake by liver cells nearly twice. The results of gene expression showed that 100 μmol/L phytanic acid significantly up-regulated glucose transporter (2.2 times), glucose transporter 2 (3 times) and glucokinase (3 times) gene expression levels, while palmitic acid only up-regulated the expression of glucose transporter 1 gene, and had no effect on the expression of glucose transporter 2 gene. Palmitic acid and DHA also showed a tendency to inhibit the expression of glucokinase gene, which suggested that phytane acid could increase glucose uptake and oxidative utilization in liver cells. It was explained in the study of Gloerich et al. (2007, 2005) that adding 0.5% of phytol into the basal diet of mice could significantly reduce plasma fatty acids, increase the free carnitine in plasma and liver and promote-oxidation of fatty acid. of which, the activities of related metabolic enzymes (acyl-coA oxidase, carnitine palmityl transferase, 3-hydroxyalkyl-coA dehydrogenase, 3-ketoyl-coA thiolase) which involved in peroxidase and mitochondrial-oxidation process and the expression levels of multiple regulatory target genes (SCOX, SCPx, DB-PHY and Catalase) of peroxisome proliferative activator receptor α (PPARα) in mice liver increased. In addition, the study of Hellgren found that after fed the mice with the diets containing 0.2% and 0.5% of phytol for 4 weeks, the liver triglyceride decreased significantly, and the phytanic acid concentration in serum and liver increased (Hellgren et al., 1999). In addition, expression regulation of key metabolizing genes of phytol in body could also have an effect on the liver glucose and lipid metabolism. Sterol carrier protein 2 (Scp2) is an essential carrier protein involved in phytanic acid transported to peroxidase. It was found in the study of Ellinghaus et al. (1999) that Scp2 gene knockout significantly increased serum phytanic acid concentration in mice, and increased the expression level of liver PPARα target genes, such as acyl-coa oxidase, peroxidase, 3-ketoyl-coA thiolase and liver fatty acid binding protein genes, etc. Above results suggested that phytol and its metabolites could reduce fat accumulation in liver by up-regulating the gene expression level of fatty acid oxidation in liver.

4.1.4 Signal pathways of phytol and its metabolites regulating glucolipid metabolism

Some studies have indicated that the regulation of phytol on adipocyte differentiation polyester and glycolipid metabolism was closely related to activation of nuclear receptors such as PPAR and retinoid X receptor (RXR).

4.1.5 Signal pathway of phytol and its metabolites regulating PPARα

It was reported in the study of Qi Lifeng et al. (2003) that PPARα was an important nuclear receptor that regulates the fatty acids oxidation, of which gene was mainly expressed in fat and liver tissues. Some studies have shown that the regulation of phytanic acid on glycolipid metabolism might be mediated by phytanic acid activation of PPARα. Heim et al. (2002) studied the activation effects of pirinic acid (WY-14643), palmitic acid, DHA, prostaglanin, cyclolitazone, phytanic acid and phytol in various configurations on PPARα in CV1 cells transfected with ACO-PPPE reporter plasmid, and found that the activation effect of three configurations phytanic acids [(3R, 7R, 11R), (3RS, 7R, 11R) and (3S, 7R, 11R)] was stronger than that of PPARα activator (WY-14643). It was also found in the the study of Ellinghaus et al. (1999) that the activation effect of phytinic acid on PPARα was about 4 times, 8 times and 9 times higher than the known PPARα agonists (bezafibrate, arachidonic acid and palmitic acid), respectively. Recent years, some studies have shown that phytol could be directly used as a ligand to activate PPARα. It was also found in the luciferase reporter gene study of Goto et al. (2005) that phytol could significantly activate PPARα, and of which activation effect was 4 times that of phytanic acid. The yeast two-hybrid test also explained that phytol could induce the binding of steroid receptor activator-1 to PPARα, which was consistent with the same concentration of fenofibrate (PPAPα agonist). In addition, HepG2 cells treated with phytol and fenofibrate could significantly up-regulate the expression of fatty acid transport protein 4 (FATP4) and a variety of PPARα target genes, such as CPT1A, ACS and ACO.

4.1.6 Signal pathway of phytol and its metabolites regulating PPARγ

The study of Garcia-rojas et al. (2005) found that after adding phytol metabolites (phytanic acid and pristanic acid) into bovine adipocytes, the phytanic acid could significantly increase the expression of PPAR γ mRNA in adipocytes, and the 100 μmol/L phytanic acid was about 6 times higher than that in control. However, 100 μmol/L pristanic acid could increase about 4.5 times. The activation effects of various terpenoids, such as geraniol, farnesol, carchine and phytol, on PPAR γ were compared in the study of Takahashi et al. (2002) by using photosynthase reporter gene. The results showed that the activation effect of 100 μmol/L phytol on PPAR γ was about twice higher than that of the control, which was similar to that of 10 μmol/L fenofibrate. The study of Alfonso et al. (2001) showed that different concentrations of phytanic acid were used to treat CHO cells transfected with PPAR γ and found that the activation of PPAR γ by 100 and 200 μmol/L phytanic acid was 2 and 3 times higher than that of the control, respectively. These results indicated that phyol and phytanic acid could activate PPAR γ, and participated in the regulation of glycolipid metabolism.

4.1.7 Signal pathway of phytol and its metabolites regulating RXR

Sutisak et al. (1996) transfected RXR CRBP ii-CAT reporter plasmid into CHO cells, and then used phytanic acid, linoleic acid, palmitic acid, arachidonic acid and 9-cis-retinoic acid to induce respectively. The results showed that the activity of CAT induced by phytanic acid was 5 times higher than that of linoleic acid, palmitic acid and arachidonic acid, and was 1/200 of 9-cis-retinoic acid. Peter et al. (1996) transfected CRBP ii-RXRE luciferase reporter gene and RXRα gene expression plasmid in SL-3 cells, and compared the activation and binding ability on RXRα among 9-cis-retinoic acid, all-trans retinoic acid and phytanic acid. The results of activation test showed that the semi-maximum effect concentrations (EC50) of 9-cis-retinoic acid, all-trans retinoic acid and phytanic acid were 2.5nmol/L, 26.0 nmol/L and 3,000.0 nmol/L, respectively. It was also found in the ligand binding assay that the semi-inhibitory concentrations (IC50) of 9-cis-retinoic acid and phytanic acid were 70.0 nmol/L, and 2,300.0 nmol/L, respectively. These results indicated that although phytanic acid could activate RXR, its activation activity was relatively weak. Some studies have also shown that phytol could affect RXR activity indirectly by regulating the metabolism of retinoic acid. The study of Alfonso et al. (2001) showed that phytol could dose-dependently inhibit the formation of all-trans retinoic acid after adding 10 μmol/L vitamin A and different concentrations of phytol (0.01 ~ 5.00 mmol/L) in human duodenal epithelial cells.

4.2 Conclusion

In conclusion, phytol and its metabolites play an important role in regulating glycolipid metabolism and polyester differentiation of adipocytes, of which mechanism was related to the activation of PPAR and RXR nuclear receptors (Lin, et al., 2012). However, the effects of phytol on lipid metabolism disorders such as treating obesity, increasing insulin sensitivity etc. need further study. In addition, activation of PPARα and PPAR γ is an important factor affecting metabolic type conversion of skeletal muscle. Therefore, further study on the effects of phytol on skeletal muscle types and meat quality of livestock, as well as the regularities of deposition and distribution in body, is not only contribute to exploring nutritional regulation measures to improve meat quality, but also of great significance to enhance the functional nutritional added value of meat.

Chapter 1 General introduction

1.1 General background

Forage is widely used by ruminants as a major feed resource for meat and milk production. Recently, the use of concentrate feeds has increased in intensive livestock production, however, the concentrate feeds are also an important source for human beings. Therefore, making flexible use of forage is an important strategy for agricultural development in the future, and exploring the potential value of forage is a necessary study.

Recently, consumers have increased concern for livestock products enriched with bioactive compounds that impact on human health. The content of such functional compounds in ruminant products is considered to relate to their content in diets consumed by animals. Abundant chlorophyll and carotenoids contained in green forage are related to the function of ruminant products for human health. In the rumen, the phytol moiety of chlorophyll is released and metabolized to phytanic acid (3, 7, 11, 15-tetramethylhexadecanoic acid) which appears into milk and meat produced by ruminants (Ackman and Hansen 1967; Wanders et al., 2011). Phytanic acid acts as one of natural ligands and activators of the peroxisome proliferator-activated receptors (PPARs) (Zomer et al., 2000; McMarty, 2001) which regulate hepatic fatty acid oxidation. In addition, some reports showed that phytanic acid had potential effects on adjusting insulin sensibility (Grimaldi, 2007) suggesting that it possesses beneficial properties for human health and prevents metabolic syndrome (Grimaldi, 2007; Hellgren, 2010). Therefore, effective utilization of the chlorophyll and phytol in forage could improve the additional value of ruminant production, and might have positive effects on the health of cows.

1.2 Photosynthetic pigments and phytol in herbages

Green forage plants contain chloroplast rich in chlorophylls, carotenoids, protein, and unsaturated fatty acids which affect the nutritional and sensory property of ruminant products. The content of such compounds in ruminant products is considered to relate to their contents in forage consumed by animals. Because the primary precursor of phytanic acid is phytol moiety of chlorophyll, phytanic acid content in ruminant products may be related to chlorophyll content in forage ingested.

Forage species, fertilization, harvest time, and conservation methods affect the forage quality (Noziere et al., 2006; Van Ranst et al., 2009; Sarmadi et al., 2016). Nitrogen (N) fertilizers can increase the content of crude protein (CP) and carotenoids (-carotene) in forages (King et al., 2012; Kopsell et al., 2007). Some reports showed that herbage harvested in early stage can have higher CP and -carotene contents (Fraser et al., 2001). However, the factors affecting the contents of chlorophyll and phytol in forages are unclear.

Silage is one of the main conserved fodders used for ruminant production (Wilkinson, 2005). Many reports have shown that -carotene is readily oxidized and that its content in forage decreases during ensiling (Bruhn and Oliver, 1978; Lindqvist et al., 2012).

However, there are few reports on the degradation of chlorophyll and other carotenoids such as lutein during ensiling.

1.3 Decomposition pathways of chlorophyll

In general, chlorophyll has two decomposition pathways: (1) under the acidic condition, the central magnesium ion of chlorophyll can be removed and converted to pheophytin, and then the pheophytin is further converted to pheophorbide and phytol; (2) chlorophyll is firstly decomposed into phytol and chlorophyllide by the action of chlorophyllase, which is subsequently converted to pheophorbide with release of magnesium ion (Matile et al., 1996). Chlorophyll is decomposed during storage, but it is unclear whether the phytol content changes in different storage conditions.

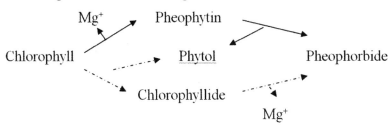

Figure 1.1 The chlorophyll decomposition pathways

1.4 Functional compounds in ruminant products

In dairy products, along with certain fatty acid (FA), carotenoids and fat-soluble vitamins are recognized as tracer compounds indicating good quality of milk and animal-feeding management (Martin et al., 2005).

The phytanic acid content in ruminant products would vary according to the chlorophyll content in forage, as well as the amount of phytol liberated in the rumen after injection of forage. Carotenoids act as antioxidants (Olson, 1989; Mangels et al., 1993) playing important roles in promoting animal and human health. Lutein is one of natural carotenoids, which widely found in vegetables, flowers, fruits, and certain algae species (Mangels et al., 1993). It is also a main component of macular pigment in the human retina. The potential benefits for lutein supplementation such as cancer prevention, enhanced immune function (Chew et al., 1996), inhibition of the autoxidation of cellular lipids (Zhang et al., 1991) and protection against oxidant-induced cell damage (Martin et al., 1996) have been reported.

Phytanic acid content in milk from the cows fed silage was found to be higher than that in milk from the cows fed hay (Markus et al., 2012). This indirectly showed that phytol content in silage was higher than that in hay. However, the change of phytol in herbage during the preservative process was unclear. In addition, the conversion ratio of phytol to phytanic acid in the rumen were also unclear. Further more, there is no report about the relationship between phytol content in diet and phytanic acid in milk.

Recently, the studies on adding carotenoid to ruminant diets have been paid more attention, and the additives have positive effects on animal health and contribute to produce animal products with additional value (Jensen et al., 1990; Kaewlamun et al., 2011; Xu et al., 2014). Several studies have shown the effects of β-carotene on reproduction, immune function and health in cows and calves (Michal et al., 1994; Kume and Toharmat, 2001).

Chapter 1　General introduction

1.5　Objectives

Italian ryegrass (Lam.), abbreviated to IR, is one of the most important forage crops. IR is now widely distributed through temperate areas of the world, and generally regarded as the basis of grassland improvement because of its high nutritional value, digestibility, and well ensiling characteristics (Breese, 1983). IR is also used as a major silage crop in Japan and has been widely used for silage making (Shao et al., 2002). Thus, in this study, IR was used as the main object to explore the changes of photosynthetic pigments and phytol and their utilizations for ruminants.

The study aimed to increase phytol content of herbage so as to produce milk containing high phytanic acid. Therefore, this study had 3 main purposes as follows:

Sub-objective 1:

Explore the factors affecting chlorophyll and phytol content in IR herbages.

Sub-objective 2:

Explore the factors affecting ruminal phytanic acid production from chlorophyll or phytol in IR herbage.

Sub-objective 3:

Explore the effect of forage sources on phytanic acid content in milk of dairy cows.

Based on the results of these studies, the changes of chlorophyll and phytol in green forage and the conversion efficiency of phytol to phytanic acid in ruminants could be confirmed. And this contributes to develop more valuable ruminant feeding methods and provide valuable theoretical basis for planting herbage and producing ruminant products in the future.

Chapter 2 Changes of photosynthetic pigments and phytol in herbages

2.1 Changes of photosynthetic pigments and phytol content at different fertilization levels in fresh herbage and hay

2.1.1 Introduction

Forage species, fertilization, harvest time, and conservation methods affect the forage quality (Noziere et al., 2006; Van Ranst et al., 2009; King et al., 2012, 2013). Nitrogen (N) fertilizers can increase the content of CP and carotenoids (-carotene) in forages (Kopsell et al., 2007; King et al., 2012). Because -carotene is easily oxidized once plants are harvested, -carotene content in stored forage is lower than that in fresh forage (Bruhn and Oliver, 1978; Kalac and McDonald, 1981; Park et al., 1983). Chlorophyll content increases with N fertilizer application (Hak et al., 1993), but gradually decreases with drying after harvest (Ballet et al., 2000). However, factors affecting chlorophyll and phytol contents in herbages, and the changes of these compounds during drying have not been studied extensively. In addition, although for the decomposition pathways of chlorophyll in aging plants have been elucidated, the changes the chlorophyll content in the whole harvested herbage are unclear. Therefore, the experiment aimed to investigate effect of N fertilizer levels on the contents of photosynthetic pigments in fresh IR herbages and their changes after drying under natural condition.

2.1.2 Materials and methods

2.1.2.1 Experiment design

The two repeated experiments (Experiment 1) were conducted in 2013 to 2014 and 2015 to 2016. In October 2013 and 2015, Italian ryegrass [Lam. variety: Ace (2013) and Inazuma (2015)] was planted in two separate fields (≥ 1 ha) at the Setouchi Field Science Center, Saijo Station, Hiroshima University (34°23′ N, 132°43′ E). In March of both years, three plots (2 × 2 m) were arranged in each field and three fertilization treatments, control: 0 kg N/ha, 60 N: 60 kg N/ha, and 120 N: 120 kg N/ha, were applied for each field (1 plots × 3 treatments × 2 fields). The herbages were harvested at heading stage on May 9th, 2014 (Variety: Ace) and April 25th, 2016 (Inazuma). The weather conditions of 2014 and 2016 were shown in Table 2.1.

A portion of the samples from harvested fresh herbage was dried in air forced oven for the proximate analysis and moisture. Another portion of the samples was kept in ice box and then preserved at -30 ℃ within 3 hours after harvesting for photosynthetic pigments and phytol analysis. The rest of harvested herbage samples (about 2 kg) were dried for 7 days under natural conditions, then a part of samples of the dried herbage (hay) were preserved at -30 ℃ for the analyses of chlorophyll and phytol.

2.1.2.2 Chemical analysis

The ground samples of the oven-dried fresh herbages and hay were analyzed for dry matter (DM), CP, crude ash and ether extract (EE) according to the method of the Association of Official Analytical

Chapter 2 Changes of photosynthetic pigments and phytol in herbages

Chemists (AOAC 1990), and to determine the neutral detergent fiber content without ash (NDFom) according to Van Soest et al. (1991). The non-fiber carbohydrate (NFC) content was calculated as follows: NFC = 100 − CP − EE − NDF − crude ash (NRC 2001).

To quantify photosynthetic pigments and phytols, freeze-dried samples (30 mg) were extracted three times with 80% acetone (Porra et al., 1989), yielding a final volume of 30 ml extract. Then, 1 mL each of trans-Apo-8'-carotenal (0.1 mg/mL acetone) and 1-nonadecanol (0.25 mg/mL hexane) were added to the extract as an internal standard for the pigment analysis and phytol analysis, respectively. Chlorophyll, chlorophyll, lutein and -carotene in the extracts were analyzed by high-performance liquid chromatography (HPLC, JASCO, Tokyo, Japan) according to the method of Zapata et al. (2000). These compounds were separated on a Symmetry-C_8 column (3.5 μm, 4.6 × 150 mm; Waters, Milford, MA, USA) at 40 °C by gradient elution. For gradient HPLC analysis, two solvents were prepared: solvent A consisted of methanol, acetonitrile, and 0.25 mol/L aqueous pyridine (50: 25: 25; pH 5.0), while solvent B consisted of methanol, acetonitrile, and acetone (20: 60: 20).

Total phytols (esterified plus free phytols) were analyzed by the methods of Liljenberg and Odham (1968) and Takeda et al. (1983). Briefly, 5 mL of the acetone extract in a screw-cap tube was dried under an N_2 gas stream at 40 °C before adding 1 mL methanol and 0.25 mL 25% (w/w) potassium hydroxide solution. Then, the mixture was heated at 70 °C for 30 min to saponify and release phytol. After heating, the mixture was extracted twice with 3 mL diethyl ether to obtain free phytol. The extracted ether phase was dehydrated with sodium sulfide and dried. The dried extract was dissolved in hexane (1 mL), loaded onto a column containing 0.1 g active carbon and 1.5 g alumina, and then eluted with 10 mL ethyl acetate. The eluate was dried at 40 °C under an N_2 stream and redissolved in 1 mL hexane for gas chromatography (GC-2014; Shimadzu, Kyoto, Japan) on a Rxi-5ms capillary column (30 m length, 0.25 mm ID, 0.25 μm df, Restec, Bellefonte, PA, USA). The analytical conditions were as follows: injector temperature, 250 ℃; detector temperature, 340 ℃; gas pressure, 90 kPa; column gas flow rate, 1.05 mL/min. The temperature program was as follows: 60 ℃ for 1 min, increase to 160 ℃ at 30 ℃/min, then to 240 ℃ at 5℃/min, and then to 320 ℃ at 20 ℃/min. Free phytol was analyzed as described above, but without the saponification step.

Phytol recovery during saponification and purification steps of the phytol analysis was tested using the commercially available standard of chlorophyll (C6144-1MG, Sigma-Aldrich, Tokyo, Japan). The chlorophyll (2 to 8 mg/L) in 80% acetone was saponified and purified by the same procedure as described above. The recovery of the phytol (107%, $R^2=0.999$) was calculated by the regression analysis between chlorophyll concentration in the standard solutions and measured phytol concentration. Chlorophyll concentration in the standard was verified by the colorimetry (E1%1 cm = 958 at 431 nm) (Lichtenthaler, 1987) using spectrophotometer (V-530, JASCO, Tokyo, Japan).

2.1.2.3 Statistical analysis

Statistical analysis was performed using the GLM procedure of Statistical Analysis System (SAS 2004). The data of chemical composition and photosynthetic pigments of fresh herbages and hay, and the ratio of hay to fresh herbages were subjected to the two-way analysis with year and fertilizer treatments as main factors. Tukey' test were used to identify differences ($P<0.05$) in treatment means between fertilization treatments at same year.

2.1.3 Results

DM yields in 2014 were 7.61, 10.01 and 10.04 t/ha for control, 60 N and 120 N treatments, respectively, while those in 2016 were 4.26, 7.07 and 7.32 t/ha for control, 60 N and 120 N treatments,

Chapter 2 Changes of photosynthetic pigments and phytol in herbages

respectively. The leaf-to-stem ratio of the fresh herbages was not affected by N fertilizer application at both years. However, the leaf-to-stem ratios in 2014 (0.53) were higher than those in 2016 (0.35).

For the fresh herbage, as the N fertilizer application rate increased, the contents of crude ash, CP, EE and NDF increased ($P<0.05$), whereas the NFC content decreased ($P<0.001$, Table 2.2). The contents of chlorophyll, phytol, lutein and β-carotene in the fresh herbage also increased ($P<0.05$) with increasing N fertilizer application rate. In addition, the contents of chlorophyll b, phytol, lutein and β-carotene of the fresh herbage harvested in 2016 were higher ($P<0.05$) than that harvested in 2014 (Table 2.2). The variation trend of chemical component and phytol content in the hay was similar to those of the fresh herbages, and was affected by N fertilizer application rate, except for β-carotene (Table 2.3). The contents of chlorophyll ($P<0.01$), lutein ($P<0.01$) and total phytol ($P<0.10$) of the hay harvested in 2016 tended to be higher than those of the hay harvested in 2014. However, there was no interaction between year and fertilizer in both the fresh herbage and hay. In addition, chlorophyll decomposition products such as pheophytin, pheophorbide and chlorophyllied were not detected in the hay.

The ratios of hay to fresh herbage for chemical components and photosynthetic pigments were shown in Table 2.4. It was obvious that the contents of chlorophyll, lutein, β-carotene and total phytol decreased by drying, and the decomposition ratio of β-carotene and lutein was about 72%~90% and 31%~69% respectively. The decomposition ratio of chlorophyll and total phytol was 40%~71% and 25%~47% respectively.

2.1.4 Discussion

The present findings confirmed that increasing the application rate of N fertilizer increased the contents of photosynthetic pigments and CP in herbages (Whitehead, 1995; Lefsrud et al., 2010). Plant photosynthetic pigments (chlorophyll and carotenoid) are mainly concentrated in the leaves. Thus, the differences in the pigment contents of the herbage in response to different fertilizer application rates probably reflected the increased pigment contents of the leaves affected by N fertilizer. In fact, no differences in the leaf-to-stem ratio (0.53, 0.52, 0.54 in 2014 and 0.32, 0.35, 0.38 in 2016 for control, 60 N and 120 N, respectively) were observed among the N fertilization treatments, even though, the averaged ratio in 2016 (0.35) was lower than that (0.53). These results indicates that the chlorophyll content in herbages is not always related to the leaf-to-stem ratio of herbages.

Although the factors affecting lutein content in forage have not been extensively investigated (Kalač 2012), the present study revealed that increasing N fertilization also increased the lutein content in forages.

The experimental results showed that the content of some chemical component and photosynthetic pigments in herbages (fresh herbage and hay) were affected by the year. The contents of EE, NDFom and lutein of fresh herbages harvested in 2016 were higher. The differences might be due to the properties of different variety used in 2014 and 2016. The IR planted in 2016 was an early maturing variety which would have higher content of these components at heading stage compared with the late maturing variety. In addition, amount of precipitation in 2016 was higher than that in 2014 (Table 2.1) and the watering rate could accelerate the growth of plant roots and promote plant growth (Endo et al., 2014) which might affect herbage chemical components. The contents of phytol and chlorophyll of the hay harvested in 2016 were higher than that harvested in 2014, even though chlorophyll content in fresh herbages was not different. Therefore, not only variety but also the drying condition might contribute the differences in the decomposition ratio of photosynthetic pigment and phytol between years.

The obvious reduction of photosynthetic pigment (chlorophyll, lutein and β-carotene) in hay should be caused by UV and high temperature (Ballet et al., 2000). Surprisingly, however,

Chapter 2 Changes of photosynthetic pigments and phytol in herbages

chlorophyll decomposition products (pheophytin, pheophorbide and chlorophyllide) were not detected in the hay. The degradation product of chlorophyll in the harvested herbage was instable and might be decomposed further. In addition, the free phytol was not found in the hay, therefore, the phytol in hay may exist as the esterified form.

The decomposition ratio of chlorophyll was higher than that of phytol at both years and all fertilization levels, because chlorophyll derivatives were decomposed further by drying, while the phytol derived from chlorophyll was presumably combined with other substances and existed as more stable esterified form.

In conclusions, the CP, EE, photosynthetic pigments and phytol in IR (fresh herbage and hay) increased with increasing N fertilization levels. Photosynthetic pigments decreased obviously by hay preparation (chlorophylls: 40%~70%, phytol: 25%~47%, β-carotene: 72%~90%, lutein: 31%~69%).

Table 2.1 Mean monthly weather conditions from March to May in 2014 and 2016 at Higashi-hiroshima

Year	Month	Temperature /°C		Rainfall /mm
		Minimum	Maximum	
2014	March	-5.5	21.4	153
	April	4.4	18.4	64
	May	10.2	23.8	71
2016	March	1.3	14.1	65
	April	7.5	20.1	208
	May	11.9	24.1	139

Data were quoted from the report by Japan Meteorological Agency.

Table 2.2 Effects of nitrogen fertilizer application rate (control, 0 kg N/ha; 60 N, 60 kg N/ha; 120 N, 120 kg N/ha) on chemical composition and photosynthetic pigment (g/kg DM) of fresh Italian ryegrass in 2014 and 2016

Item	2014			2016			SEM	P-value		
	Control	60N	120N	Control	60N	120N		N	Y	N*Y
Moisture (g/kg FM)	749	806	833	782	825	849	50.8	<0.001	0.001	0.378
Crude ash	56.8	72.7	80.4	55.5	74.1	87.2	3.65	0.001	0.437	0.503
Crude protein	50.6	78.9	103.2	61.1	74.1	128.3	13.41	0.015	0.379	0.561
Ether extract	15.3	21.7	24.9	20.8	29.7	33.6	2.97	0.029	0.026	0.853
NDFom	413	488	500	414	537	541	17.8	0.001	0.036	0.316
NFC	406	268	224	445	285	210	17.2	<0.001	0.364	0.391
Chlorophyll a	1.38	2.45	2.87	1.50	2.30	4.20	0.508	0.015	0.296	0.307
Chlorophyll b	0.50	0.91	1.24	1.03	1.68	1.52	0.273	0.096	0.043	0.628
Chlorophyll a+b	1.88	3.35	4.11	2.52	3.97	5.73	0.747	0.026	0.155	0.676
Total phytol	0.78	1.55	1.78	1.22	2.21	2.71	0.074	<0.001	0.001	0.054
Lutein	0.286	0.466	0.496	0.420	0.616	1.21	0.1083	0.010	0.009	0.053
β-carotene	0.099	0.169	0.185	0.118	0.190	0.451	0.0630	0.034	0.047	0.130

SEM, standard error of the mean; NDFom, neutral detergent fiber exclusive of residual ash; NFC, non-fibrous carbohydrate; N, effect of nitrogen fertilizer application rate ;Y, effect of year; N*Y, interaction effect between N and Y.

Chapter 2　Changes of photosynthetic pigments and phytol in herbages

Table 2.3 Effects of nitrogen fertilizer application rate (control, 0 kg N/ha; 60N, 60 kg N/ha; 120N, 120 kg N/ha) on chemical composition and photosynthetic pigment (g/kg DM) of Italian ryegrass hay in 2014 and 2016

Item	2014			2016			SEM	P-value		
	Control	60N	120N	Control	60N	120N		N	Y	N*Y
Moisture (g/kg FM)	158	134	119	134	121	109	13.9	0.122	0.183	0.845
Crude ash	54.7	71.6	78.6	57.7	83.8	80.7	5.751	0.017	0.272	0.647
Crude protein	44.2	59.5	79.8	58.4	73.7	96.3	8.253	0.018	0.076	0.988
Ether extract	12.6	14.9	18.4	16.1	19.9	21.4	1.92	0.083	0.059	0.871
NDFom	421	509	521	497	557	567	21.1	0.001	0.001	0.342
NFC	468	345	302	341	263	221	25.0	0.003	0.006	0.931
Chlorophyll a	0.47	0.69	1.11	0.66	1.39	2.27	0.274	0.025	0.029	0.294
Chlorophyll b	0.14	0.33	0.64	0.75	0.86	0.94	0.091	0.039	0.002	0.337
Chlorophyll a+b	0.61	1.03	1.75	1.41	2.26	3.21	0.299	0.012	0.005	0.537
Total phytol	0.49	0.99	0.92	0.91	1.40	1.66	0.148	0.021	0.077	0.503
Lutein	0.110	0.146	0.243	0.240	0.427	0.661	0.0737	0.024	0.004	0.196
β-carotene	0.013	0.017	0.031	0.027	0.045	0.097	0.0201	0.180	0.083	0.481

SEM, standard error of the mean; NDFom, neutral detergent fiber exclusive of residual ash; NFC, ous carbohydrate; N, effect of nitrogen fertilizer application rate ; Y, effect of year;N*Y, interaction effect between N and Y.

Table 2.4 Effects of nitrogen fertilizer application rate (control, 0 kg N/ha; 60N, 60 kg N/ha; 120N, 120 kg N/ha) on the ratio of hay to fresh herbages for chemical component in 2014 and 2016

Item	2014			2016			SEM	P-value		
	Control	60N	120N	Control	60N	120N		N	Y	N*Y
Chlorophyll a	0.34	0.28	0.39	0.43	0.62	0.54	0.047	0.243	0.002	0.072
Chlorophyll b	0.34	0.37	0.52	0.72	0.56	0.65	0.083	0.307	0.010	0.242
Chlorophyll a+b	0.34	0.29	0.43	0.56	0.60	0.57	0.005	0.451	0.002	0.253
Total phytol	0.62	0.66	0.53	0.75	0.63	0.61	0.118	0.622	0.547	0.807
Lutein	0.39a	0.31b	0.49a	0.58	0.69	0.55	0.038	0.636	0.001	0.022
Carotene	0.13	0.10	0.18	0.24	0.24	0.28	0.105	0.876	0.226	0.975

a,b Means with different letters significantly differ (P<0.05).
SEM, standard error of the mean; NDFom, neutral detergent fiber exclusive of residual ash; NFC, non-fibrous carbohydrate; N, effect of nitrogen fertilizer application rate ;Y, effect of year; N*Y, interaction effect between N and Y.

2.2 Changes in carotenoid, chlorophyll and phytol contents in Italian ryegrass during ensiling

2.2.1 Introduction

In Experiment 1, it was found that the photosynthetic pigments and phytol in harvested herbage decreased as the herbage was dried under natural conditions, where the herbage was exposed in air and UV. However, it was unclear that whether the herbage sealed and kept in dark place could contribute to the preservation of photosynthetic pigments and phytol. Grass silage is one of the main conserved fodders used for ruminant production (Wilkinson 2005). Many reports have shown that -carotene is readily oxidized and that its content in forage decreases during ensiling (Bruhn and Oliver 1978; Lindqvist et al., 2012). However, there are few reports on the degradation of chlorophyll and other carotenoids such as lutein during ensiling. Although the decomposition pathways of chlorophyll of ageing plants were indicated, those in silage were not reported. The degradation of carotenoids and chlorophylls during ensiling may affect their final contents and phytol content in silage. In addition, the extent of the degradation of these components during ensiling remains unclear.

Therefore, the objective of this study was to investigate the changes in the contents of carotenoids and chlorophylls, and their derivatives during the ensiling of IR.

2.2.2 Materials and methods

2.2.2.1. Experimental materials and ensiling procedures

Italian ryegrass [Lam. (Variety: Ace)] was planted in the experimental field (4.2 ha) at the Setouchi Field Science Center,

Saijo Station, Hiroshima University, for a winter growing season in 2013 to 2014. In the experimental field, the herbage was harvested at the heading stage in 17 May 2014, and allowed to wilt under natural conditions for 1 day to reach a moisture content of approximately 60%. The weeds in the wilted herbage were removed before preparing silage. One portion each of the fresh harvested herbage sample and wilted herbage sample was stored at -30 ℃, and another portion was dried at 70 ℃ for proximate analysis. Table 2.5 shows the chemical composition of the fresh herbage.

The wilted herbage was ensiled using a small-scale pouch system. Approximately 400 g wilted herbage was cut into 2 cm pieces, which was packed into each plastic film bag (Hiryu BN-11 type 370 mm × 250 mm; Asahikasei, Tokyo, Japan). Twelve bags were prepared in total, which were sealed with a vacuum sealer (SQ303W; Sharp, Osaka, Japan). Every three bags were unsealed at 1st, 2nd, 3rd and 5th week, respectively. After the bags were opened, cold-water extracts were prepared for pH analysis according to Cai et al. (1999). The pH of the water extract was measured with a glass electrode pH meter (Horiba F-72, Horiba, Kyoto, Japan). Moreover, the extracts were preserved at -30 ℃ for further analysis. Sub-samples of these silages were lyophilized for further analysis.

2.2.2.2 Chemical analysis

The ground dried samples of fresh herbage, wilted herbage and silages were analyzed to determine the contents of DM, CP, crude ash and EE according to the AOAC (1990), and NDFom according to Van Soest et al. (1991). The NFC content was calculated as follows: NFC = 100 − CP − EE − NDF − crude ash.

The analysis methods of pheophorbide, pheophytin, pheophytin, chlorophyll, chlorophyll, lutein, -carotene and phytol content were same as Experiment 1. The defrost water extract (1 mL) was mixed with 0.2 mL metaphosphoric acid (25% w/w), then centrifuged at 10,000 for 10 min at 4 ℃. The obtained supernatant added with 4-metyl valeric acid as an internal standard was analyzed for volatile fatty acids (VFA) by gas chromatography (GC-17A; Shimadzu, Kyoto, Japan) with a column (CP-FFAP CB, 25 m lengths, 0.32 mm ID, 0.3 μm df, Agilent Technologies, California, USA) by the following conditions: injector temperature, 200 ℃; detector temperature, 250 ℃; gas pressure 60 kPa. Initial column temperature was hold at 80 ℃ for 2 min, then increased to 120 ℃ at 10 ℃/min, then further increased to 130 ℃ at 1 ℃/min, and finally increased to 240 ℃ at 30 ℃/min. Lactic acid contents in the water extracts were measured by a test kit (1112821, Roche/R-Biopharm, Darmstadt, Germany) with a micro plate reader (Multiscan GO; Thermo Fisher Scientific Inc., Massachusetts, USA).

2.2.2.3 Statistical analysis

Statistical analysis was performed using the GLM procedure of Statistical Analysis System (SAS, 2004). The content of non-detectable compounds was regarded as zero for statistical analyses. The chemical composition and photosynthetic pigment data were subjected to one-way analysis of variance. Tukey' test were used to identify differences ($P<0.05$) in the contents of components over time.

2.2.3 Results

Although moisture content of the pre-ensiled herbage (53%) was lower than the planned level (60%) after wilting for one day, the moisture content in silage did not change over the 5-week period (Table 2.6). Even in the first week, lactic acid, acetic acid, iso-butyric acid and butyric acid were detected in silage. Lactic, acetic and butyric acid showed the highest content ($P<0.05$) at week 5. The initial pH of the water extract of material herbage (6.46) was higher than that of silage at first week. Reflecting the change of organic acid content over the ensiling period, pH of silage was gradually decreased ($P<0.05$) from week 2 to week 5 (Table 2.6).

During the ensiling process in the 5-weeks, the contents of CP and EE increased ($P<0.05$), while the NDFom and NFC contents did not change significantly (Table 2.7). The chlorophyll content decreased dramatically in the first week ($P<0.05$), and remained almost constant from week 2 to week 5, while chlorophyll was not detected after week 2 (Table 2.8). The pheophytin content increased during the 5-week ensiling period, while the pheophytin content showed very low values in the first two weeks, and higher ($P<0.05$) values in weeks 3 and 5. The pheophorbide content increased continuously over the 5-week ensiling period. The molar content of phytol was higher than that of total chlorophyll (chlorophyll +) at the start of the ensiling process (week 0). Although the molar content of total chlorophylldecreased, that of total phytol did not change significantly over the 5-week period. Free phytol was barely detectable during the ensiling period. Instead, most phytol was in the esterified form. Among the carotenoids, the lutein content in silage did not change until week 5. The-carotene content decreased gradually during first two weeks ($P<0.05$), but did not change significantly after that.

2.2.4 Discussion

This study evaluated the chlorophyll and carotenoid contents in IR before, during and after 5-weeks ensiling, to determine the changes in the contents of these functional components in low moisture silage. In general, moisture content of wrapped roll-bale grass silage is recommended to reduce by 50 to 60% by wilting for preparing good quality silage (Beaulian et al., 1993). It is assumed that the reduction of moisture content of herbage also reduce the chlorophyll degradation during ensiling due to less production of organic acid and relatively higher pH of low moisture silage (Uchida et al., 1989; Beaulian et al., 1993). The initial reduction of pH is expected to be occurred at during first 2 weeks (Zhu et al., 1999), thus this experiment focused on the changes in the pigment and carotenoid contents during first 5 weeks of ensiling.

The CP content of herbage used in this study was relatively low, due to late harvest stage. Many reports have shown that the CP content increases after ensiling (Nishino and Uchida, 1999), consistent with this results. The results of this experiment showed that the EE content increased over time during ensiling, possibly because of the esterification of various compounds in the herbage during the ensiling process.

During ensiling, the contents of chlorophyll, chlorophyll and -carotene decreased, consistent with the results of other reports (Makoni and Shelford, 1993; Noziere et al., 2006). The chlorophyll and chlorophyll contents decreased sharply at the beginning of the ensiling process, but did not change significantly after week 2 during ensiling. Compared with the initial contents of total chlorophyll and -carotene, the proportions remaining after the 5-week ensiling process were 20% and 75%, respectively. Although the lutein content showed slight fluctuations, it did not change significantly during ensiling, suggesting that lutein in IR is not readily decomposed during this process.

Because pheophytin and are derived from chlorophyll and, respectively the changes in pheophytin and contents showed opposite trends to those of chlorophyll and contents during ensiling. Because chlorophyll is decomposed to pheophytin under acidic conditions, the changes in chlorophyll and pheophytin contents at the first week may reflect the organic acids production and reduction of pH during this period. Although pH value decreased gradually with the increasing ensiling time, and the content of acetic acid, butyric acid and lactic acid increased obviously at week 5, microbial fermentation seemed to proceed gradually during ensiling due to low moisture content in herbage.

The results showed that a substantial amount of chlorophyll decomposed within the first week of ensiling, while the pheophytin content began to increase after week 2 of ensiling. Some previous literatures reported that, on the process of chlorophyll metabolism, some of chlorophyll was transformed into chlorophyll (Vezitskii, 2000). At the early stage of decomposition, thus, some chlorophyll may be transformed into chlorophyll which is then degraded via the chlorophyll decomposition pathway (Hörtensteiner and Kräutler 2000). Because pheophorbide is derived from pheophytin, the pheophorbide content increased as the pheophytin content increased.

In the experiment, chlorophyllide was not detected in this experiment, implying that most of the chlorophyll was decomposed into pheophytin, which then released phytol (Hoyt 1970). Based on results, the following pathways of chlorophyll degradation during ensiling was proposed: some chlorophyll was transformed into pheophytin which was transformed into pheophorbide, releasing phytol; the rest of the chlorophyll was transformed into chlorophyll which decomposed to form pheophytin which transformed into pheophorbide, releasing phytol.

In this experiment, the contents of both esterified phytol and free phytols in silage were measured. Interestingly, free phytol was barely detected, even though the formation of pheophorbide from pheophytin should release free phytol. It would be responsible for re-esterification of released phytol under the weak acidic conditions during ensiling. Some reports have pointed out that, in senescent leaves, released phytol is re-esterified and combines with other substances (Csupor 1971; Peisker et al., 1989). Moreover, the molar content of phytol was higher than that of chlorophylls in the herbage material. Therefore, some of the phytol may be derived from substances other than chlorophyll, for example, long-chain fatty acid esters or phylloquinone (vitamin K_1)(Peisker et al., 1989).

In conclusion, the -carotene content decreased during ensiling, but the lutein content did not. Although the chlorophyll content decreased by 80%, and the chlorophylls were gradually converted into pheophytin and pheophorbide during the 5-week ensiling process, the total phytol content did not change significantly. The re-esterification

Chapter 2 Changes of photosynthetic pigments and phytol in herbages

of free phytol released from chlorophyll and phytol derived from compounds other than chlorophyll may have contributed to the stable phytol content in IR silage.

Table 2.5 Chemical composition (g/kg DM) of fresh herbage of Italian ryegrass

Item	Content
Moisture (g/kg FM)	862
Crude ash	66.1
Crude protein	56.0
Ether extract	14.7
Neutral detergent fiber	516
Non-fibrous carbohydrate	347

Table 2.6 Changes of fermentation characteristics (g/kg DM) during ensiling process

Item	Ensiling period (weeks)				SEM
	1	2	3	5	
Moisture (g/kg FM)	53.8	53.9	53.6	53.0	0.010
pH	6.32^a	6.23^a	6.12^b	5.91^c	0.001
Lactic acid	0.24^b	0.24^b	0.25^b	0.31^a	0.005
Acetic acid	3.86^b	3.87^b	4.26^b	4.84^a	0.025
Propionic acid	ND	ND	ND	0.27^a	0.005
Isobutyric acid	0.57	0.56	0.67	0.44	0.012
Butyric acid	0.30^c	0.30^c	0.34^b	0.43^a	0.001

^{a,b,c} Means with different letters significantly differ ($P<0.05$).
SEM: standard error of the means.

Table 2.7 Changes in chemical composition (g/kg DM) of Italian ryegrass herbage during ensiling

Item	Ensiling period (weeks)					SEM
	0	1	2	3	5	
Crude ash	66.3^c	70.1^{bc}	71.6^{ab}	73.7^{ab}	75.5^a	0.21
Crude protein	56.3^b	54.7^b	56.8^b	61.4^{ab}	64.5^a	1.53

续表

Item	Ensiling period (weeks)					SEM
	0	1	2	3	5	
Ether extract	14.8^c	17.8^b	21.0^a	22.2^a	22.7^a	0.52
NDFom	514	518	505	509	516	9.30
NFC	347	340	346	333	322	10.1

^{a,b,c}Means with different letters significantly differ ($P<0.05$).
SEM: standard error of the means. NDFom: neutral detergent fiber exclusive of residual ash; NFC, non-fibrous carbohydrate

Table 2.8 Changes in photosynthetic pigment contents (mmol/kg DM) of Italian ryegrass herbage during ensiling

Item	Ensiling period (weeks)					SEM
	0	1	2	3	5	
Chlorophyll a	1.32^a	0.49^b	0.39^b	0.35^b	0.38^b	0.038
Chlorophyll b	0.60^a	0.44^b	ND	ND	ND	0.023
Chlorophyll a+b	1.91^a	0.93^b	0.39^c	0.35^c	0.38^c	0.048
Pheophytin a	0.021^c	0.405^b	0.510^{ab}	0.524^a	0.591^a	0.021
Pheophytin b	ND	ND	0.012^c	0.112^b	0.209^a	0.018
Pheophorbide a	ND	0.674^c	0.892^b	0.962^b	1.317^a	0.041
Lutein	0.354	0.424	0.320	0.421	0.469	0.057
β-carotene	0.123^a	0.114^{ab}	0.081^b	0.089^b	0.091^b	0.005
Total phytol	2.41	2.61	2.77	2.77	3.04	0.592

^{a,b,c}Means with different letters significantly differ ($P<0.05$).
SEM: standard error of the means.

2.3 Effects of nitrogen fertilization and harvesting stage on photosynthetic pigments and phytol contents of Italian ryegrass silage

2.3.1 Introduction

It was confirmed, in the Experiment 1, that N fertilizer was an effective way to improve the contents of chlorophyll and phytol. In Experiment 2, phytol and lutein were found to be preserved well during ensiling. In addition, the photosynthetic pigments and chemical component are also affected by harvesting stages (Fraser et al., 2001; Zhao et al., 2003; Spanghero et al., 2015). Therefore it is hypothesised that the content of photosynthetic pigment in ensiled herbage also affected by harvesting stages and N fertilization levels.

Therefore, the objectives of the present experiment were to investigate the effects of N fertilizer application and harvesting stage on the contents of chlorophyll, phytol, and carotenoids in pre-ensiled herbage and silage of IR (Lolium multiflorum Lam.) and to determine the extent of phytol decomposition after ensiling.

2.3.2. Materials and methods

2.3.2.1 Experimental design

Italian ryegrass [Lam. (Variety: Ace)] was planted at the Setouchi Field Science Center, Saijo Station, Hiroshima University (34°23′ N, 132°43′ E) in October 2013 in three separate fields (≥ 1 ha). Immediately prior to sowing, manure (compost composed of fermented dung and sawdust shavings from a dairy cowshed) was applied on two fields at the rates of 32 t/ha and 33 t/ha, respectively, and chemical fertilizer (urea) was applied at the rates of 54 kg N/ha and 55 kg N/ha, respectively. No manure was applied on the third field and only a compound fertilizer (N: P: K = 14: 14: 14) was applied at the rate of 49 kg N/ha. In late March 2014, an experimental block (4 × 6 m, 24 m^2) was established on each field and each block was further divided into six plots of 4 m^2 each. On March 29, three N fertilizer treatments (control: 0 kg N/ha; 60 N: 60 kg N/ha; and 120 N: 120 kg N/ha) were applied, by topdressing with urea, in each block (2 plots × 3 treatments × 3 blocks). The herbage in one of the two plots with the same fertilizer treatment in each block (1 plot × 3 treatments × 3 blocks) was harvested at the booting stage (April 25, 2014, 27 weeks of age), and the herbage in the remaining plots was harvested at the heading stage (May 9, 2014, 29 weeks of age). The herbage was cut 3 cm above the ground, and total weight of the fresh herbage in each plot was measured. Approximately 200 g of fresh herbage from each plot was separated into leaves and stems, then dried at 60 °C for 2 days in a forced-air oven to measure DM content and DM ratio of stem to leaf in the fresh herbages. The monthly maximum and minimum temperatures and monthly rainfall recorded at the study site (Higashihiroshima-shi) in March, April, and May 2014 were 21.4 °C, −5.5 °C, and 153 mm, 18.4 °C, 4.4 °C, and 64 mm, and 23.8 °C, 10.2 °C, and 71 mm, respectively.

2.3.2.2 Silage preparation

Harvested herbage from each plot was ensiled in small silage bags (Hiryu BN-11 type, 370 mm × 250 mm; Asahikasei, Tokyo, Japan). The herbage harvested from each plot was wilted for 1 day then any weeds were removed before preparing for ensiling. The wilted herbage from each plot (400 g each) was chopped into 2 cm pieces and packed into individual plastic film bags. The samples of the wilted herbages collected immediately before ensiling (pre-ensiled herbages) were stored at -30°C for later analysis. The bags containing the wilted herbage were sealed using a vacuum sealer (SQ303W; Sharp, Osaka, Japan), then stored in a room at 20~25 °C for 60 days. At 60 days of

Chapter 2 Changes of photosynthetic pigments and phytol in herbages

ensiling, the bags were opened, and part of the silage sample was stored at -30 ℃. Cold-water extracts of silage were also prepared to measure pH and organic acid content in silage as described by Cai et al. (1999). The pH of the water extract was measured using a glass electrode pH meter (Horiba F-72, Horiba, Kyoto, Japan). For further analyses, the extracts were stored at -30 ℃. The frozen samples of both the pre-ensiled herbages and silages were lyophilized.

2.3.2.3 Chemical analysis

The dried samples of both the pre-ensiled herbages and silages were analyzed to determine the contents of DM, CP, crude ash, and EE according to the methods of the Association of AOAC (1990) and the NDFom as described by Van Soest et al. (1991). The NFC and total carbohydrate contents were calculated as follows: NFC = 100 − CP − EE − NDFom − crude ash (NRC 2001); total carbohydrate = NFC + NDFom.

The analysis methods of pheophorbide, pheophytin, pheophytin, chlorophyll, chlorophyll, lutein, -carotene, phytol content were the same as Experiment 1. And the analysis methods of VFA and lactic acid were the same as Experiment 2.

2.3.2.4 Statistical analysis

As small bag silages were prepared with herbages obtained from each plot separately, one plot was regarded as an experimental unit and three blocks were planned to be replicates for the randomized block design (RBD). Because the type of basal fertilization was different among the blocks established on the different fields, the interactions between block and fertilization levels or harvesting stage were tested by two-way ANOVA, as a first step. As a result, the interactions were not detected for all parameters. Thus, this experiment was regarded as a 3 × 2 factorial arrangement in a RBD (3 replicates (blocks) × 3 fertilizer treatments × 2 harvesting stages, $n=18$). The statistical analysis was conducted using the MIXED procedure of SAS (SAS

Institute, 2004) with the following model: = $\mu + \alpha + + (\alpha) + \gamma+$, where is the observed value, μ is the overall mean, α is the effect of N fertilizer (i=1-3), is the effect of harvesting stage (j=1-2), α is the interaction between fertilizer and harvesting stage, γ is the random effect of blocks (k=1-3), and is random error. When a significant interaction between the effect of N fertilizer and that of harvesting stage was detected, Tukey-Kramer test was used to identify the significant difference in treatment means between harvesting stages at same fertilization levels. Significance was declared at P<0.05 and a tendency was considered up to P < 0.10.

2.3.3 Results

2.3.3.1 Dry matter yield of herbages

At the booting stage, the DM yields of fresh herbages for the control, 60N and 120N treatments were 3.51, 4.67 and 5.13 t/ha, respectively. At the heading stage, the DM yields for the control, 60N and 120N treatments were 6.18, 8.05 and 8.28 t/ha, respectively. Although the difference in DM yield between the 60N and 120N treatments was small, N fertilizer significantly increased DM yield (P<0.05). In addition, DM yield at the heading stage was higher (P<0.01) than that at the booting stage. The leaf-to-stem ratio of the fresh herbages was not affected by N fertilizer application at both harvest stages (data not shown). However, the leaf-to-stem ratios at the booting stage (1.7) were significantly higher (P<0.05) than those at the heading stage (0.8).

2.3.3.2 Effect of nitrogen fertilizer application on chemical composition of herbages

As the N fertilizer application rate increased, the contents of moisture, CP, EE and NDFom in both the pre-ensiled herbage and silage increased (P<0.05), whereas the NFC and total carbohydrate contents decreased (P<0.01)(Table 2.9). The contents of chlorophyll,

Chapter 2　Changes of photosynthetic pigments and phytol in herbages

chlorophyll, phytol, lutein and -carotene in the pre-ensiled herbage increased ($P<0.05$) with increasing N fertilizer application rate (Table 2.10). In the silage, the contents of chlorophyll, lutein and -carotene increased ($P<0.05$) with increasing N fertilizer application rate, whereas no chlorophyll was detected. The contents of pheophorbide, pheophytin andpheophytin in the silage increased ($P<0.05$) as the N fertilizer application rate increased (Table 2.10). The total phytol contents in the pre-ensiled herbage and silage increased ($P<0.01$) with increasing N fertilizer application rate. The pH and contents of lactic, acetic, propionic, and butyric acids in the silages were not affected by N fertilizer application (Table 2.11).

2.3.3.3 Effect of harvesting stage on chemical composition in herbages

The contents of moisture, crude ash, CP, and EE in the pre-ensiled herbage harvested at the booting stage were higher ($P<0.05$) than those at the heading stage (Table 2.9). The content of NDFom in the pre-ensiled herbage at the booting stage was lower ($P<0.01$) than that at the heading stage. The NFC content in the pre-ensiled herbage was similar at both the booting and heading stages, while the total carbohydrate content was higher ($P<0.01$) at the heading stage (Table 2.9). In the silage, differences in the contents of these compounds between the booting and heading stages showed similar tendencies to those of the pre-ensiled herbage. The contents of chlorophylls, lutein, and -carotene in the pre-ensiled herbage at the booting stage were higher ($P<0.05$) than those at the heading stage (Table 2.10). In the silage, the contents of pheophytin and lutein at the booting stage were higher ($P<0.05$) than those at the heading stage, whereas the contents of chlorophyll, pheophorbidepheophytin, and -carotene showed no significant differences between the booting and heading stages. The total phytol contents were higher at the booting stage than those at the heading stage in the pre-ensiled herbage ($P<0.05$) and the silage ($P<0.01$). The pH of the silage at the booting stage was lower ($P<0.05$)

than that at the heading stage (Table 2.11). In silage at the heading stage, the lactic acid content was lower ($P<0.05$) and the butyric acid content tended to be lower ($P=0.05$), however, no propionic acid was detected (Table 2.11).

2.3.3.4 Interaction between nitrogen fertilization and harvesting stage on chemical composition of herbages

The moisture and CP contents of the pre-ensiled herbage and silage were significantly affected by the interaction of fertilizer application rate and harvesting stage (Table 2.9). Neither the moisture nor CP contents were significantly different between the booting and heading stages for the control, while these were higher ($P<0.05$) at the booting stage compared with those at heading stage for the 60N and 120N treatments. In the silage, the content of pheophytin in the control did not differ significantly between the booting and heading stages (Table 2.10). However, at the 60 N and 120 N fertilizer application rates, pheophytin content at the booting stage was significantly higher ($P<0.05$) than that at the heading stage. No significant interaction effect of the two factors was observed for silage fermentation characters (Table 2.11).

2.3.4 Discussion

The relationship between the contents of CP and photosynthetic pigments and N fertilization levels has been discussed in Experiment 1. The carbohydrate content in herbages was also affected by the fertilization levels. Although reports on the effect of N fertilization levels on the fiber content of herbages have been inconsistent, some studies (Keady and O'Kiely, 1996; Masoero et al., 2011; Heeren et al., 2014) reported that increasing N fertilization levels decreased the water-soluble carbohydrate content and increased the fiber content of herbages as observed in the present study. However, in the present study, changes in the carbohydrate fraction with the fertilization levels

Chapter 2　Changes of photosynthetic pigments and phytol in herbages

had little effect on the fermentation quality of the silage.

In the silage, the contents of CP, carotenoids and chlorophyll-related compounds other than chlorophyll reflected changes in the respective contents in the pre-ensiled herbage with the N fertilization rate. Although the chlorophyll content in the pre-ensiled herbage increased with increasing N fertilization levels, no chlorophyll was detected in the silage. This finding may have been caused by the extensive decomposition of chlorophyll to pheophytin during ensiling.

Changes in the nutrient and pigment contents with harvesting stage were also observed in the present study. The harvesting stage affects the chemical composition and physiological property of forages (Buxton, 1996); forage harvested at an early developmental stage contains higher CP and EE, and lower NDF than those at an advanced stage (King et al., 2012). The interaction between fertilization levels and harvesting stage observed for the CP contents indicated that the decrease in CP content in herbage with advancing maturity was affected by the rate of N fertilizer application. Although the reasons for this interactive effect are not clear, the relatively smaller change in the total carbohydrate content between the stages for the control fertilizer treatment (Table 2.9) may have induced a non-significant reduction in the CP content at the heading stage. The present results also indicated that silage made from herbage harvested at an early stage is an effective means of increasing the supply of lutein, phytol, CP, and fat to animals, though the DM yields at an early stage is relatively low. In the present study, the average DM yields of forage at the booting and heading stages were 4.44 and 7.48 t/ha, respectively.

Although the content of chlorophyll in the pre-ensiled herbages differed between the harvest stages, its content in the silages did not (Table 2.10) because the chlorophyll decomposed during ensiling. The ratio of chlorophyll in the silage to pre-ensiled herbage shown in Table 2.12 indicates the proportion of chlorophyll remaining in herbage after ensiling. It was numerically higher at the heading stage than that at the booting stage for the control and 60N treatments, whereas it was higher at the booting stage for the 120N treatment. This implies that the extent of chlorophyll degradation during ensiling may be affected by the harvest stage and rate of N fertilizer application.

In the present study, all the harvested fresh forages were wilted under natural conditions for 1 day. As the temperature in April (booting stage) was lower than that in May (heading stage), the moisture content of the pre-ensiled herbage and silage was higher at the booting stage. This may have facilitated fermentation at the booting stage compared with that at the heading stage, because higher moisture content could enhance the activity of lactic acid bacteria and contribute to the generation of lactic acid (Garcia et al., 1989). In the present study, the mean pH of the silage was lower and the organic acid content was higher at the booting stage, compared with that at the heading stage. The lower pH value of silage at the booting stage might have affected the extent of chlorophyll decomposition because the rate of chlorophyll degradation in plant materials depends on the pH during preservation (Koca et al., 2006). However, there was no relationship between the pH or organic acid content and the phytol content in the silage in the present study.

As expressed by molar basis, the phytol content was higher than that of chlorophyll at both harvest stages with all fertilization levels (Table 2.12). It was confirmed that the phytol may be derived from substances other than chlorophyll.

A large proportion (80%~85%) of chlorophyll in the herbage degraded to pheophytin and pheophorbide, but the phytol was well preserved (87%~130%) during ensiling for all N fertilization treatments at both harvest stages (Table 2.12). This high recovery of phytol in the silage indicates that ensiling is an effective method for preserving phytol in herbage. No free phytol was detected in the silage, even though the conversion of pheophytin to pheophorbide releases phytol. This indicated that the phytol derived from pheophytin may be esterified again with a different substance, such as a fatty acid, and that this esterified phytol is likely to be more resistant to

Chapter 2 Changes of photosynthetic pigments and phytol in herbages

decomposition. However, phytol content in hay dried naturally for 7 days decreased by 25%~47% as indicated in Experiment 1. Previously, it has been reported that dairy cows fed silage exhibited a higher plasma concentration of phytanic acid (Verhoeven et al., 1998), suggesting that phytol may be better preserved in silage. In addition, a regression analysis revealed that the phytol content in silage was linearly correlated ($r=0.97$, $P<0.001$) with the CP content (Figure 2.1). Thus, the phytol content in silage can be estimated indirectly from the CP content.

β-carotene could be affected by the quality of silage (Kalač and Kyzlink, 1979; Lindqvist et al., 2012). However, the relatively higher pH of the silage in the present study, ranging from 5.1 to 5.9, might have contributed to the extensive preservation of β-carotene during ensiling. Lutein content generally decreases after the plant material is dried (Reynoso et al., 2004). However, in the present study, the ratio of β-carotene and lutein in the silage to that in the pre-ensiled herbage ranged from 0.82 to 1.66 and 0.77 to 1.58, respectively (Table 2.12). Therefore, ensilage could restrain the decomposition of lutein as well as β-carotene.

In conclusion, the IR silage harvested at early stage or under high N fertilization could have high content of CP, EE and photosynthetic pigment related compounds. After ensiling, phytol was higher in silage harvested at early stage or under high N fertilization. In addition, part of phytol may come from the decomposition of other substances besides chlorophyll. Lutein and phytol in IR could be preserved well during ensiling.

Table 2.9 Effects of nitrogen fertilizer application rate (control, 0 kg N/ha; 60N, 60 kg N/ha; 120N, 120 kg N/ha) and harvest stage (booting and heading stages) on chemical composition (g/kg DM) of Italian ryegrass herbage and silage

Item	Nitrogen fertilizer application rate						SEM	P-value		
	Control		60N		120N			N	H	N*H
	Booting	Heading	Booting	Heading	Booting	Heading				
Herbage (g/kg DM)										
Moisture (g/kg FM)	588	559	690a	569b	710a	564b	49.0	0.004	<0.001	0.010
Crude ash	58.8	54.6	74.3	62.5	84.3	65.9	11.31	0.023	0.030	0.464
Crude protein	60.7	45.8	99.6a	59.5b	134.0a	79.2b	8.41	<0.001	<0.001	0.020
Ether extract	19.5	15.8	26.0	19.7	31.2	21.1	1.62	<0.001	<0.001	0.071
NDFom	350	411	415	482	426	495	24.9	0.003	0.001	0.984
NFC	512	472	385	376	338	339	38.7	<0.001	0.411	0.652
Total carbohydrate	862	883	800	858	764	834	17.5	<0.001	<0.001	0.117
Silage (g/kg DM)										
Moisture (g/kg FM)	617	543	726a	575b	730a	554b	42.3	0.001	<0.001	0.008
Crude ash	74.8	64.9	90.7	71.9	101.0	76.6	13.93	0.064	0.012	0.603
Crude protein	82.1	63.3	120.0a	75.3b	145.0a	97.1b	6.07	<0.001	<0.001	0.016
Ether extract	27.2	22.0	39.6	24.4	44.1	27.9	2.49	0.001	<0.001	0.059
NDFom	400	452	482	515	503	524	33.1	0.004	0.065	0.756

Chapter 2 Changes of photosynthetic pigments and phytol in herbages

Item		Nitrogen fertilizer application rate							SEM	P-value		
		Control		60N		120N				N	H	N*H
		Booting	Heading	Booting	Heading	Booting	Heading					
NFC		416	398	267	314	208	275		49.9	0.001	0.209	0.342
Total carbohydrate		816	850	749	829	711	799		18.9	<0.001	<0.001	0.080

a,b Means within same fertilizer application rate marked with different superscripts differ significantly (P<0.05).
SEM, standard error of the mean; DM, dry matter; FM, fresh matter; NDFom, neutral detergent fiber exclusive of residual ash; NFC, non-fibrous carbohydrate; N, effect of nitrogen fertilizer application rate; H, effect of harvest stage; N*H, interaction effect between N and H.

Table 2.10 Effects of nitrogen fertilizer application rate (control, 0 kg N/ha; 60N, 60 kg N/ha; 120N, 120 kg N/ha) and harvest stage (booting and heading stages) on photosynthetic pigment contents (g/kg DM) of Italian ryegrass herbage and silage

Item	Nitrogen fertilizer application rate						SEM	P-value		
	Control		60N		120N			N	H	N*H
	Booting	Heading	Booting	Heading	Booting	Heading				
Herbage (g/kg DM)										
Chlorophyll a	1.22	0.93	2.55	1.60	3.39	2.49	0.272	<0.001	0.002	0.259
Chlorophyll b	0.44	0.39	0.98	0.71	1.43	1.15	0.097	<0.001	0.024	0.402
Chlorophyll a+b	1.66	1.31	3.53	2.31	4.82	3.65	0.387	<0.001	0.002	0.239
Total phytol	0.87	0.70	1.50	1.38	2.02	1.56	0.141	<0.001	0.035	0.378
Lutein	0.427	0.153	0.753	0.288	1.003	0.440	0.0568	<0.001	<0.001	0.061
β-carotene	0.093	0.060	0.178	0.096	0.230	0.163	0.0145	<0.001	<0.001	0.200
Silage (g/kg DM)										
Chlorophyll a	0.273	0.260	0.613	0.450	0.750	0.527	0.1039	0.006	0.101	0.512
Chlorophyll b	ND	ND	ND	ND	ND	ND	-	-	-	-
Pheophytin a	0.413	0.346	0.893a	0.567b	1.203a	0.740b	0.0844	<0.001	<0.001	0.022
Pheophytin b	0.103	0.127	0.207	0.190	0.290	0.243	0.0271	<0.001	0.536	0.420
Pheophorbide a	0.963	0.686	1.160	1.140	2.520	1.536	0.3819	0.013	0.151	0.372
Total phytol	1.11	0.73	1.93	1.19	3.43	1.64	0.163	<0.001	<0.001	0.060

Chapter 2 Changes of photosynthetic pigments and phytol in herbages

Item	Nitrogen fertilizer application rate							P-value		
	Control		60N		120N		SEM	N	H	N*H
	Booting	Heading	Booting	Heading	Booting	Heading				
Lutein	0.370	0.246	0.560	0.413	0.833	0.473	0.0582	<0.001	0.001	0.123
β-carotene	0.100	0.097	0.143	0.113	0.240	0.160	0.0316	0.025	0.174	0.494

a,b,c Means within same fertilizer application rate marked with different superscripts letter differ significantly ($P<0.05$).
SEM, standard error of the mean; N, effect of nitrogen fertilizer application rate; H, effect of harvest stage; N*H, interaction effect between N and H.

Table 2.11 Effects of nitrogen fertilizer application rate (control, 0 kg N/ha; 60N, 60 kg N/ha; 120N, 120 kg N/ha) and harvest stage (booting and heading stages) on fermentation characteristics (g/kg DM) of Italian ryegrass silage

Item	Nitrogen fertilizer application rate						SEM	P-value		
	Control		60N		120N			N	H	N*H
	Booting	Heading	Booting	Heading	Booting	Heading				
pH	5.54	5.64	5.10	5.76	5.62	5.86	0.17	0.040	0.270	0.293
Lactic acid	2.68	0.63	3.45	1.22	3.93	1.11	0.95	0.481	0.004	0.858
Acetic acid	5.79	6.24	8.98	4.71	3.04	4.41	1.40	0.139	0.507	0.189
Propionic acid	0.17	ND	ND	ND	0.37	ND	–	–	–	–
Butyric acid	1.58	0.51	8.22	0.47	1.51	0.29	2.37	0.427	0.051	0.419

SEM, standard error of the mean; N, effect of nitrogen fertilizer application rate; H, effect of harvest stage; N*H, interaction effect between N and H; ND, not detected.

Chapter 2 Changes of photosynthetic pigments and phytol in herbages

Table 2.12 Effects of nitrogen fertilizer application rate (control, 0 kg N/ha; 60N, 60 kg N/ha; 120N, 120 kg N/ha) and harvest stage (booting and heading stages) on chlorophyll and phytol contents (mmol/kg DM) in Italian ryegrass herbage and silage

Item	Nitrogen fertilizer application rate						SEM	P-value			
	Control		60N		120N			N	H	N*H	
	Booting	Heading	Booting	Heading	Booting	Heading					
Herbage (mmol/kg DM)											
Chlorophyll a+b	1.85	1.47	3.94	2.57	5.37	4.06	0.332	<0.001	0.002	0.239	
Phytol	2.93	2.36	5.06	4.65	6.82	5.26	0.472	<0.001	0.033	0.364	
Phytol/Chlorophyll a+b1	1.60	1.62	1.33	1.86	1.29	1.31	0.201	0.271	0.277	0.381	
Silage (mmol/kg DM)											
Chlorophyll a+b	0.308	0.292	0.686	0.497	0.838	0.588	0.116	0.006	0.093	0.503	
Phytol	3.70	2.50	6.50	4.00	9.50	5.53	0.055	<0.001	<0.001	0.060	
Silage to herbage ratio2											
Chlorophyll a+b	0.171	0.196	0.179	0.196	0.154	0.145	0.028	0.356	0.640	0.819	
Phytol	1.33	1.05	1.29	0.87	1.40	1.04	0.040	0.566	0.011	0.881	
Lutein	0.87	1.58	0.77	1.56	0.82	1.10	0.227	0.489	0.010	0.487	
β-carotene	1.09	1.66	0.82	1.19	1.05	1.00	0.237	0.156	0.261	0.450	

SEM, standard error of the mean; N, effect of nitrogen fertilizer application rate; H, effect of harvest stage; N*H, interaction effect between N and H.

1 The molar ratio of phytol to chlorophyll in herbage.
2 The proportion of chlorophyll or phytol that remained after ensiling.

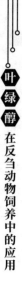
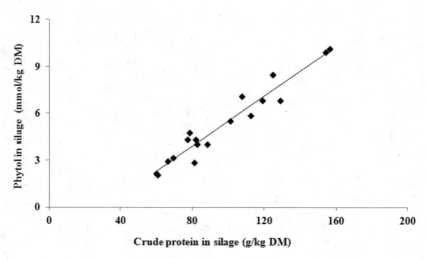

Figure 2.1 Relationship between crude protein content (X) and total phytol content (Y) in silage. $Y=0.080X-2.51$; $r^2 =0.94$, $P<0.001$

2.4 Effects of lactic acid bacteria addition on photosynthetic pigments of ensiled Italian ryegrass grown under different nitrogen fertilization levels

2.4.1 Introduction

The chlorophyll and phytol in the harvested herbage decomposed by drying under natural condition (Experiment 1), while phytol in IR could be preserved well during ensiling (Experiments 2 and 3). In addition, silage pH has been shown to affect the preservation of -carotene (Noziere et al., 2006). However, it is unclear whether the silage pH affects the chlorophyll or phytol contents in herbage after ensiling. Also, few reports presented the effects of silage additives on the decomposition of chlorophyll and phytol during ensiling.

Changes in photosynthetic pigment of herbages at different fertilization levels during ensiling were confirmed in Experiments 2 and 3. However, the fermentation level of the silage was relative low, as the pH value was relatively high and organic acid contents were low. Therefore, different fermentation levels during ensiling may affect on photosynthetic pigment decomposition and preservation in silage.

Lactic acid bacteria (LAB) are generally used to enhance the fermentation of silage and prevent the growth of fungi or yeast that could cause aerobic spoilage (Woolford, 1990). The pH of silage decreased dramatically and the generation of lactic acid was accelerated after adding LAB (Tian et al., 2014). However, it was unclear that the effects of adding LAB on photosynthetic pigments of silage at different fertilization levels.

Therefore, the experiment aimed to explore the effects of adding LAB on photosynthetic pigments and phytol in ensiled IR grown at different fertilizer levels.

2.4.2 Materials and methods

2.4.2.1 Experimental design

Italian ryegrass [Lam. (Variety: Ace)] was planted in three separate fields (≥ 1 ha) at the Setouchi Field Science Center, Saijo Station, Hiroshima University (34°23′ N, 132°43′ E). In March 2015, three plots (2 × 2 m) were arranged in each field, then three fertilization treatments, control: 0 kg N/ha, 60 N: 60 kg N/ha, and 120 N: 120 kg N/ha, were applied for each field. The herbages were harvested at heading stage on April 29th, 2015, and allowed to wilt under natural conditions for 1 day to reach a moisture content of approximately 65%. The weeds in the wilted herbage were removed before preparing silage. Prior to ensiling, two different treatments: (1) without additive, (2) addition of lactic acid bacteria (LAB: 5 mg/kg fresh grass, Si-master LP spray, Snow Brand Seed, Sapporo, Japan) were applied for wilted herbages at three different fertilization levels.

The wilted herbage was ensiled using a small-scale pouch system. Approximately 400 g wilted herbage was cut into 2 cm pieces and packed with the additive into each plastic film bags (Hiryu BN-11 type 370 mm × 250 mm; Asahikasei, Tokyo, Japan). Subsequently bags were sealed with a vacuum sealer (SQ303W; Sharp, Osaka, Japan), and stored in a room at 20 %~25 ℃ for 60 days. After the bags were opened, cold-water extracts were prepared for pH analysis according to Cai et al. (1999). The pH of the extract was measured by a glass electrode pH meter (Horiba F-72; Horiba, Kyoto, Japan). For further analysis, the extracts were preserved at -30 ℃ and sub-samples of these silages were lyophilized.

2.4.2.2 Chemical analysis

The ground dried samples of pre-ensiled herbages and silages were analyzed to determine the contents of DM, CP, crude ash and EE and NDFom and the analysis methods were the same as Experiment 1. The NFC content was calculated as follows: NFC = 100 – CP – EE – NDF – crude ash (NRC 2001). The analysis methods of pheophorbide, pheophytin, pheophytin, chlorophyll, chlorophyll, lutein, -carotene, phytol contents, VFA and lactic acid contents were the same as Experiment 2.

2.4.2.3 Statistical analysis

Statistical analysis were performed using the GLM procedure of Statistical Analysis System (SAS 2004). The content of non-detectable compounds was regarded as zero for statistical analyses. The chemical composition and photosynthetic pigment of pre-ensilage herbage were subjected to one-way analysis of variance. Tukey' test was used to identify the differences ($P<0.05$) in treatment means between fertilizer levels. The chemical composition, photosynthetic pigment and fermentative characteristics data of silage were subjected to two-way analysis of variance.

2.4.3 Results

The DM yields of fresh herbages for the control, 60N and 120N treatments were 5.54, 6.26 and 6.45 t/ha, respectively. Although the difference in DM yield between the 60N and 120N treatments was small, N fertilizer significantly increased DM yield ($P<0.05$).

As the N fertilizer application rate increased, the contents of CP, EE, NDFom, chlorophyll a, phytol, lutein and β-carotene in the pre-ensiled herbage increased ($P<0.05$, Table 2.13). There were no obvious differences in NDF and lutein content in pre-ensiled herbages between 60N and 120N (Table 2.13). CP, EE, NDF contents in silage had the same tendency as pre-ensiled herbages whose content increased with the increase of N fertilization. Pheophytin a, pheophytin b, pheophorbide a, total phytol and lutein contents in silage increased with the increase of N fertilization ($P<0.01$, Table 2.14). However, chlorophyll b was not detected in silage (Table 2.14). The content of chlorophyll a, pheophytin a, pheophytin b and total phytol content in the silage were not affected by the LAB addition. However, pheophorbide a content was obviously lower for the LAB treatment ($P<0.001$). In contrast, β-carotene content for the LAB treatment was obviously higher than those for the control treatment ($P<0.05$). For the fermentation quality, pH value was not affected by the N fertilization. Irrespective of N fertilization levels, pH value of silage was lower ($P<0.001$) for the LAB treatment compared with the control. Lactic acid content was extremely higher ($P<0.001$) for the LAB treatment. The contents of acetic acid, propionic acid and butyric acid were not affected by the LAB addition, although propionic acid content was lower ($P<0.05$) for the high N fertilization level (Table 2.15).

2.4.4 Discussion

The effects of N fertilizer on chemical component and

photosynthetic pigment were discussed in detail in Experiment 1 and Experiment 3, and the results of the experiment showed the similar effects of N fertilization as above experiments.

Chlorophyll was not detected after ensiling, and about 80 % chlorophyll was decomposed irrespective of the additive treatments, as observed in Experiment 3. Therefore, the decomposition ratio of chlorophyll was not affected by fertilization levels or fermentation treatments. pH value of silage was not affected by N fertilization treatment, which was inconsistent with the results of King et al, (2012), even though the contents of lactic acid and acetic acid were affected by N fertilization.

Although fermentation level and pH were different between the additive treatments, the phytol content in silage and the esterified status were not affected by the additives. Therefore, it is suggested that different fermentation quality would not affect the preservation of phytol during ensiling.

In silage, pheophorbideincreased with increase N fertilization, and lower for the LAB treatment. The results of this experiment showed that pheophorbide content in silage had correlation with pH value of the silage. Although pheophorbide mainly come from pheophytin, there were no differences in pheophytin content between the Control and LAB treatments. Therefore, a part of pheophorbide might be decomposed through the acidic condition or by the action of LAB. -carotene content in silage was almost the same among the three different fertilization levels, although high fertilization level increased -carotene content in pre-ensiled herbage. About 62%~75% -carotene was decomposed during ensiling in the control treatment, which values were slightly higher than those reported in previous study (about 40%~60%, Noziere et al., 2006).

Moreover, -carotene content in silage was affected by the additives, in which the decomposition ratio during ensiling was only 0~44% for the LAB treatment. Those result suggested that -carotene was preserved well in the low pH silage. Silage preparation at low pH

Chapter 2 Changes of photosynthetic pigments and phytol in herbages

could control the decomposition of -carotene (Noziere et al., 2006), which was also verified by the LAB treatment in this experiment. Lutein content in silage was not affected by the additives and its content did not change during ensiling, which was also consistent with the results in Experiments 2 and 3.

In conclusion, the IR under high N fertilization could have high content of CP and photosynthetic pigments. After ensiling, the phytol and lutein content was not affected by additive treatments, and could be preserved well in silage. The decomposition ratio of -carotene was lower for the LAB treatment, and could be preserved well in silage at lower pH value.

Table 2.13 Effects of nitrogen fertilizer application rate (0 N, 0 kg N/ha; 60 N, 60 kg N/ha; 120 N, 120 kg N/ha) on chemical composition and photosynthetic pigment (g/kg DM) of pre-ensiled herbage

Item	Nitrogen fertilization level (N)			SEM
	0 N	60 N	120 N	
Chemical composition (g/kg DM)				
Moisture (g/kg FM)	649	687	688	5.8
Crude ash	68.1^c	85.4^b	93.2^a	12.2
Crude protein	61.1^c	96.2^b	116.4^a	11.2
Ether extract	19.6^b	20.3^b	22.2^a	3.95
NDFom	494^b	568^a	588^a	0.3
NFC	358^a	238^{ab}	194^b	7.6
Photosynthetic pigment (g/kg DM)				
Chlorophyl a	1.90^b	2.19^{ab}	2.85^a	0.16
Chlorophyl b	0.62	0.94	0.99	0.06
Total phytol	0.91^c	1.67^b	2.54^a	2.783
Lutein	0.352^b	0.698^a	0.923^a	0.0912
β-carotene	0.077^c	0.104^b	0.146^a	0.5824

^{a,b,c} Means with different letters significantly differ ($P<0.05$).
SEM, standard error of the means; DM, dry matter; NDF, neutral detergent fiber; NFC, non-fibrous carbohydrate =100-crude protein-ether extract-neutral detergent fiber-crude ash.

Table 2.14 Effects of nitrogen fertilizer application rate (0N, 0 kg N/ha; 60N, 60 kg N/ha; 120N, 120 kg N/ha) and additives on chemical composition and photosynthetic pigments (g/kg DM) of Italian ryegrass silage

Item	0N		60N		120N		SEM	P-value		
	Control	LAB	Control	LAB	Control	LAB		N	T	N*T
Chemical composition (g/kg DM)										
Moisture (g/kg FM)	622	633	665	682	656	656	13.9	0.019	0.422	0.817
Crude ash	78.7	67.9	90.1	82.5	98.6	107.1	8.82	<0.001	0.816	0.207
Crude protein	70.1	63.9	103	105	140	138	5.72	<0.001	0.906	0.933
Ether extract	24.3	22.6	29.5	28.4	31.1	31.5	1.12	<0.001	0.388	0.530
NDFom	468	475	525	532	534	550	17.3	0.005	0.514	0.954
NFC	385	362	304	306	295	244	20.8	<0.001	0.134	0.382
Photosynthetic pigment (g/kg DM)										
Chlorophyll a	0.849	0.362	0.555	0.585	0.809	0.871	0.1213	0.106	0.220	0.088
Chlorophyll b	ND	ND	ND	ND	ND	ND				
Pheophytin a	0.735	0.581	0.848	0.928	1.061	1.278	0.0911	<0.001	0.506	0.139
Pheophytin b	0.168	0.007	0.250	0.371	0.385	0.486	0.0523	<0.001	0.317	0.088
Pheophorbide a	0.965	0.343	0.967	0.478	2.030	0.738	0.1631	0.002	<0.001	0.067
Total phytol	1.180	1.130	2.040	1.790	2.523	2.456	0.2021	<0.001	0.454	0.863

Chapter 2 Changes of photosynthetic pigments and phytol in herbages

Item	Nitrogen fertilizer application rate						SEM	P-value		
	0N		60N		120N			N	T	N*T
	Control	LAB	Control	LAB	Control	LAB				
Lutein	0.427	0.349	0.665	0.622	0.915	0.824	0.0734	<0.001	0.263	0.946
β-carotene	0.030	0.078	0.038	0.075	0.037	0.082	0.0212	0.949	0.016	0.950

SEM, standard error of the mean; N, effect of nitrogen fertilizer application rate; T, effect of additives treatment; N*T, interaction effect between N and T.

Table 2.15 Effects of nitrogen fertilizer application rate (0N, 0 kg N/ha; 60N, 60 kg N/ha; 120N, 120 kg N/ha) and additives on fermentation characteristics (g/kg DM) of Italian ryegrass silage

Item	Nitrogen fertilizer application rate						SEM	P-value		
	0N		60N		120N			N	T	N*T
	Control	LAB	Control	LAB	Control	LAB				
pH	5.59	3.89	5.71	3.57	5.98	3.97	0.111	0.125	<0.001	0.393
Lactic acid	1.74	12.2	3.57	19.4	4.21	16.6	1.152	0.003	<0.001	0.065
Acetic acid	5.39	6.16	4.53	5.81	4.24	5.38	0.873	0.144	0.533	0.540
Propionic acid	0.890	0.717	0.563	0.481	0.479	0.458	0.1113	0.020	0.305	0.768
Butyric acid	1.29	1.43	1.47	1.07	0.72	0.39	0.513	0.264	0.650	0.849

SEM, standard error of the mean; N, effect of nitrogen fertilizer application rate; T, effect of additive treatment; N*T, interaction effect between N and T.

2.5 Changes of chlorophyll and phytol contents in different harvesting height King grass before and after ensiling

2.5.1 General background

The tropical regions of China have the advantages of low latitude, strong light, various forage species and large biomass, which contribute to the production of forage with high chlorophyll content. King grass is abundant in tropical regions of China and is also an important source roughage source for tropical ruminants, which is of great significance to the development of tropical animal husbandry. The mean rainfall of Hainan Island is high, which contributes to the rapid growth of king grass, therefore, the king grass can be cut 8 ~ 10 times a year (being cut at about 2 m height each time). It was difficult to confirm the nutrient value changes of king grass through its growth period, so many studies reflected the maturity and nutritional changes through the height of king grass (Dong et al., 2013; Nurfeta et al., 2008). Previous studies have shown that chlorophyll content in grass was affected by forage species, fertilization levels and cutting stages, etc. In addition, temperature, sunlight and moisture, etc had effects on the decomposition of chlorophyll. The experiments in 2.2 and 2.3 showed that the contents of chlorophyll and phytol in the ryegrass dried under nature condition (about 70 % moisture) did not decrease significantly, which indicated that higher moisture could effectively lower the decomposition of chlorophyll. So far, there is no study on the changes of phytochrome in tropical herbage before and after silage. Therefore, it is of great significance to confirm the changes of phytochrome in king grass for the production of animal products in tropical regions. The aims of the study were to investigate the changes

Chapter 2 Changes of photosynthetic pigments and phytol in herbages

of chlorophyll and phytol in king grass (being cut at different heights) before and after silage, and to analyze the relation between the contents of chlorophyll and phytol and fermentation indexes and basic nutrient contents in king grass silage.

2.5.2 Materials and methods

2.5.2.1 King grass culture and samples

King grass (Reyan No.4) was grown in the cultivation field in the experimental base affiliated to Tropical Crop Variety Resources Institute, Chinese Academy of Tropical Agricultural Sciences (195° 109.5'N, 109° 30'E, 149 m), which was cultured under natural conditions and irrigated with adequate water. The experiment field was divided into 4 parts averagely, of which straight-line distance between each part was more than 15 m, and each part consists of four plots on average (16 plots in total). The grass (stubble height 5 cm) from one of the four plots was harvested on February 28th, 2018 (grass height 40 ~ 80 cm, T1), March 16th, 2018 (grass height 80 ~ 120 cm, T2), March $29^{th,}$ 2018 (grass height 120 ~ 180 cm, T3) and April $23^{rd,}$ 2018 (grass height 180 ~ 220 cm, T4) respectively. Firstly, the harvested king grass was weighted to calculated the dry matter (DM) biomass, and part of the samples was selected to determine the stem/leaf ratio. Then, the rest of the samples were cut to about 2 cm and mixed, then dried for 1 day under natural conditions until about 80% moisture. About 100 g of the semi-dried sample was stored at -20 ℃ for the analysis of chlorophyll and phytol content, and another 100 g sample was dried at 65 ℃ for the determination of nutrient content.

2.5.2.2 King grass silage preparation

In each plot, 200 g of the semi-dried sample was put into a polyethylene silage bag (30 cm × 20 cm), duplicated twice in each plot (32 bags in total), vacuumized with vacuum baler (Sinbo, Shanghai) and sealed, then stored in a dark room (room temperature

25~30 ℃) and fermented for 60 days before unsealed. After unsealed, about 50 g of the sample was stored at -20 ℃ for the analysis of chloro-phyll and phytol contents, and another 50 g of sample was dried at 65 ℃ for the determination of nutrient contents.

2.5.3 Sample analysis

2.5.3.1 Determination of fermentation quality of king grass silage

After unseal the silage, 200 mL distilled water was added above a 50 g of chopped sample, sealed and stored in a refrigerator (4 ℃) for 24 h, then the liquid was filtered with 4 layer of gauze. The pH of filtered fluid was measured by PHS-3C precision pH meter. Then the filtered fluid was poured into the centrifuge tube and centrifuged by HERM-LE (USA) at a speed of 12 000 r/min. After centrifugated for 5 min, a 0.22 μm microporous membrane was used to filter the liquid into the sample bottle, and the contents of lactic acid (LA), ethyl acid (AA), propionic acid (PA) and butyric acid (BA) were analyzed by high performance liquid chromatography (HPLC). HPLC setting conditions were as follows: HPLC column was RP-18 column (5 μm, 4.6 mm × 25 mm), the detector was Hitachi Primaide UV detector, the mobile phase was methanol at a flow rate of 1 mL/min, the detection wavelength was set at 210 nm and the injection volume was 10 μL. The content of volatile base nitrogen (VBN) in silage was analyzed by micro diffusion method. A diffusion dish was prepared, 1 mL boric acid indicator (containing boric acid, methyl red, bromocresol green ethanol solution) was added to the inner dish, then 1 mL centrifuged silage fluid and saturated potassium carbonate solution (K_2CO_3, 52.5%) was added to the outer dish and sealed, mixed fully and maintained in a volatile state for 24 h, finally the hydrochloric acid was titrated (0.01 mol/L).

Chapter 2　Changes of photosynthetic pigments and phytol in herbages

2.5.3.2 Determination of nutrient content

The contents of DM, crude protein (CP), ether extract (EE) and ash (Ash) in naturally dried king grass and silage king grass were determined according to the method of AOAC (1999). The content of neutral detergent fiber (NDF) was determined according to the method of Van Soest et al., the calculation of non-fiber carbohydrate (NFC) content was according to the method of NRC (2001), and the calculation formula was as follows: NFC = 100−CP−Ash−EE−NDF.

2.5.3.3 Determination of chlorophyll content

According to the method of Porra et al., the chlorophyll content in king grass before and after silage was determined as follows: The frozen sample was taken out and freeze-dried for 72 h with a freeze-dryer (Labconco, USA). After fully ground, about 50 mg of the sample was put into a 50 mL centrifuge tube, and 30 mL of 80% acetone aqueous solution was added, sealed and stored in cold storage for 24 h (in a dark room). The absorbance was determined by Spectrophotometer (Shimadzu, UV 2600, Japan) at wavelength 663.6 and 646.6 nm (recorded as A663.6 and A 646.6, respectively) to calculate the contents of chlorophyll and chlorophyll, and the sum of which was the content of chlorophyll. In order to reflect the relation between chlorophyll and phytol and explore whether all phytol molecules come from chlorophyll molecules, the unit of chlorophyll and phytol content was converted into mole content for comparison, and the calculation method is as follows:

Chlorophyll a (g/kg DM) = [(A663.6 × 12.25−2.25 × A646.6) × 30/sample weight (mg)/(1,000 × 1,000)]/[DM content (%)/100]; Chlorophyll b (g/kg DM) = [(A646.6 × 20.31−4.91 × A663.6) × 30/sample weight (mg)/(1,000 × 1,000)]/[DM content (%)/100].

2.5.3.4 Determination of phytol content

According to the method of Lv et al. (2017), the content of phytol in king grass before and after silage was determined as follows: 1 mL of 1-nonadecyl alcohol internal standard fluid (0.25 mg/mL) was added into the 30 mL extraction solution, then mixed completely for the extraction of phytol. Gas chromatograph (GC-7890A; Agilent Techonlogies, USA) with Rxi-5 capillary tube column (length 30 m, inner diameter 0.25 mm, film thickness 0.25 mm; Restec, USA) was used to analyse, and the analysis conditions were as follows: the injector temperature was 250 ℃, the detector temperature was 340 ℃, the gas pressure was 90 kPa, and the column gas flow was 1.5 mL/min. The temperature process was as follows: 60 ℃ lasts for 1 min, then increases to 160 ℃ at the rate of 30 ℃/min, then increases to 240 ℃ at the rate of 5 ℃/min, and finally increases to 320 ℃ at the rate of 20 ℃/min.

2.5.4 Statistical analysis

SAS 9.2 software was used for statistical analysis of the experimental data. The nutrient contents, chlorophyll, phytol and fermentation indexes of the king grass before and after silage were analyzed by one-way analysis of variance, with $P<0.05$ as the difference significance standard. Excel 2013 was used to analyze the relation between the contents of chlorophyll and phytol and fermentation indexes and the nutrient contents in king grass silage.

2.5.5 Results and analysis

2.5.5.1 Comparison of DM biomass and stem/leaf ratio of king grass at different cutting heights

Table 2.16 showed that the DM biomass of king grass in T1 to T4 increased significantly with the increase of cutting heights ($P<0.05$), and the ratio of stem/leaf was 1.37, 0.85, 0.69 and 0.46, respectively, which significantly decreased with the increase of cutting height ($P<0.05$).

2.5.5.2 Comparison of nutrient contents, chlorophyll and phytol of king grass at different cutting heights

The changes of nutrient contents, chlorophyll and phytol in king grass at different cutting heights were shown in Table 2.17. The contents of CP and ash in T1 to T4 were significantly decreased with the increase of cutting height ($P<0.05$), while the contents of NDF and NFC were significantly increased with the increase of cutting height ($P<0.05$). EE content in T1 was significantly higher than that in other treatments ($P<0.05$), while there was no significant difference between T2 and T3 ($P > 0.05$). In semi-dried king grass (about 80% moisture content), the contents of chlorophyll and chlorophyll were significantly decreased with the increase of cutting height ($P<0.05$), and the contents of chlorophyll and phytol showed the same trend.

2.5.5.3 Comparison of changes of nutrient contents, chlorophyll and phytol in king grass silage at different cutting heights

Table 2.18 showed the changes of nutrient contents, chlorophyll and phytol in silage at different cutting heights. After 60 days of silage, the contents of CP, NDF, EE and Ash in king grass silage showed the same trend as those in semi-dried king grass (about 80% moisture). In addition, most of the chlorophyll was decomposed after silage, and its content was significantly lower than that before silage, however the residual chlorophyll was affected by cutting height, and the contents of chlorophyll, chlorophyll and chlorophyll content in silage were significantly decreased with the increase of cutting height ($P<0.05$). Significant changes of phytol were not found in silage which was 3.04, 1.75, 1.62 and 1.36 g/kg DM at four cutting heights, respectively, and the phytol content decreased significantly with the increase of cutting height ($P<0.05$), and the changes of which in silage were the same as that of semi-dried king grass.

2.5.5.4 Comparison of chlorophyll and phytol before and after silage of king grass at different cutting heights

The molar contents of chlorophyll and phytol were calculated based on the contents of chlorophyll and phytol in king grass and king grass silage, which was shown in Table 2.19. After silage, the decomposition rates of chlorophyll, chlorophyll and chlorophyll were about 70%, 90% and 74%, respectively at different cutting heights. However, in T1 to T4, the ratios of phytol content before and after silage was 0.97, 0.92, 1.7 and 1.15, respectively, and there was no significant differences among treatments ($P>0.05$), which proved that phytol could be well preserved in silage.

2.5.5.5 Fermentation indexes of king grass silage at different cutting heights

Table 2.20 showed the fermentation indexes of king grass silage at different cutting heights. There was no significant difference in lactic acid among T1, T2 and T4 ($P>0.05$), however, the acetic acid in T1 was significantly higher than that in T2, T3 and T4 ($P<0.05$). Highest lactic acid (3.46 g/kg DM), lowest acetic acid (0.61 g/kg DM) and pH (3.62) were produced in T3 which showed excellent fermentation quality. Further analysis of the data showed that there was a significant positive correlation between phytol and CP in king grass silage ($R_2 = 0.74$, Fig.2), and there was no statistical correlation between chlorophyll and phytol and fermentation indexes (pH, volatile fatty acids and volatile base nitrogen content) ($R_2<0.4$, Table 2.20).

2.5.6 Discussion

King grass as an important feed crop in Hainan province has the characteristics of rapid growth, large yield and easy to ensile. According to the biomass data of this experiment, the chlorophyll content of the king grass at four different cutting heights was 52.1,

Chapter 2　Changes of photosynthetic pigments and phytol in herbages

57.2, 63.1 and 50.7 kg/ha DM, respectively. Although the chlorophyll content in T1 was higher, the chlorophyll content in T3 was the highest, in addition, the contents of CP and NDF in T3 were also higher. Based on the comprehensive evaluation of basic nutrient contents, chlorophyll content and DM biomass, it was found that the king grass cut at the height of 120 ~ 180 cm (T3) had higher feeding value. The chlorophyll content in T4 was the lowest and it was mainly synthesized by nitrogen and phosphorus in soil. However, the nitrogen in soil was limited, and there might be insufficient nitrogen for further chlorophyll synthesis when the king grass height was above 180 cm. Therefore, it is necessary to explore the chlorophyll content combined the soil nutrient and light data for further study.

The results showed that the cutting height had a significant effect on the contents of basic nutrient contents, chlorophyll and phytol. The chlorophyll content in ryegrass was tested in the experiments 2.1 ~ 2.4, however, the content of chlorophyll in ryegrass was significantly lower than that in king grass regardless of being cut at early stage (grass height about 40 cm) or late stage (grass height about 70 ~ 80 cm). Although the cultivation conditions were different, it also indirectly proves that the low latitude environment in tropical regions was conducive to chlorophyll synthesis. The results showed that the content of chlorophyll in king grass was more than twice greater than that in ryegrass under the same cutting height condition. With the increase of cutting height, the chlorophyll content in king grass decreased gradually. Compared with T1, the contents of chlorophyll and phytol in T4 decreased more than twice, which indicated that the cutting height had an obvious effect on the contents of chlorophyll and phytol. Some studies showed that phytochrome mainly exists in plant leaves. In the study, the stem-to-leaf ratio (leaf/stem) of king grass cut at different heights was tested on the basis of dry matter, which indirectly reflected the changes of chlorophyll content in king grass (Table 2.16), that is, with the increase of cutting height, stem-to-leaf ratio decreased gradually, and the chlorophyll content in individual king grass was diluted. After 60 days of silage, all fermentation bags were not broken. With the increase of cutting height, the changes of conventional nutrients in king grass silage was the same as that in semi-dried king grass (about 80% moisture), however, the contents of CP and NDF in king grass silage were lower than that in semi-dried king grass (about 80% moisture), which might be due to higher moisture and the loss of some nutrients along with the king grass fermentation. In the silage of king grass cut at different heights, most of the chlorophyll was decomposed and the remained chlorophyll was also affected by cutting heights, that is, the chlorophyll content decreased significantly with the increase of cutting height. In the silage process of T1 to T4, the decomposition rate of chlorophyll was almost the same (Table 2.19). After calculated, it was found that about 74% of chlorophyll in king grass silage was decomposed and about 80% of chlorophyll in ryegrass silage was decomposed (compare 2.2 with 2.3), suggesting that the forage species had effects on the decomposition of chlorophyll. In addition, the results showed that a little chlorophyll existed in the king grass silage and almost no chlorophyll was found in ryegrass silage. The content of phytol in king grass silage was similar to that in semi-dried king grass silage (about 80% moisture) (Table 2.18 and Table 2.19), indicating that phytol was well preserved in the king grass silage. Also, a linear relationship existed between phytol content and CP content ($y=0.024, 9x-2.222 4, R_2=0.74$; FIG.2), therefore, the content of phytol could be indirectly calculated based on the content of CP in king grass silage, which was the same as previous research results.

  Several studies have shown that different cutting heights had effects on the fermentation quality of forage silage (Neylon et al., 2003; Johnson et al., 2003). The results of this study showed that different cutting heights of king grass resulted in different NFC contents, which contribute to different organic acid contents in silage. After compared among the four treatments, it was found that in T3 the LA content was highest and acetic acid content was lowest, which showed the best fermentation quality (Table 2.20). Although different cutting heights resulted in significant differences in fermentation quality (volatile fatty acids and volatile base nitrogen), no statistical relationship was found among the content changes, phytol preservation and chlorophyll decomposition after statistical analysis. Therefore, fermentation quality might have no effects on the changes of chlorophyll during silage.

2.5.7 Conclusion

The contents of chlorophyll and phytol decreased significantly with the increase of cutting height. After silage, about 74% of chlorophyll was decomposed; however, phytol was well preserved. In addition, there was no significant correlation between the contents of chlorophyll and phytol and fermentation quality, however, the phytol content was positively correlated with CP content.

Table 2.16 Comparison of DM biomass and leaf-to-stem ratio of king grass at different cutting heights

Items	Cutting height				SEM
	T1	T2	T3	T4	
DM biomass/ (t/ha)	6.2^d	9.9^c	13.6^b	19.7^a	0.16
Leaf-to-stem ratio	1.37^a	0.85^b	0.69^c	0.46^d	0.09

Values in the same row with different small letter superscripts mean significant difference ($P<0.05$), while with no or the same letter superscripts mean no significant difference ($P>0.05$). The same is as below.

Table 2.17 Comparison of the contents of common nutrients, chlorophyll and phytol in king grass at different cutting heights

Items	Cutting height				SEM
	T1	T2	T3	T4	
Basic nutrient contents					
Moisture/ (g/kg FM)	79.9	80.8	81.8	78.7	0.57
CP/ (g/kg DM)	211^a	192^b	176^c	130^d	1.15
NDF/ (g/kg DM)	520^d	547^c	565^b	607^a	1.22
EE/ (g/kg DM)	28.0^a	25.1^b	25.6^b	22.2^c	0.09
Ash/ (g/kg DM)	149^a	125^b	108^c	79^d	0.78
NFC/ (g/kg DM)	92^d	111^c	125^b	162^a	1.11
Chlorophyll and phytol					
Chlorophyll a/ (g/kg DM)	6.35^a	4.47^b	3.68^c	2.31^d	0.46
Chlorophyll b/ (g/kg DM)	2.11^a	1.31^b	0.93^c	0.72^d	0.24
Chlorophyll a+b/ (g/kg DM)	8.46^a	5.78^b	4.62^c	3.03^d	0.38
Phytol/ (g/kg DM)	3.12^a	1.90^b	1.52^c	1.18^d	0.11

Table 2.18 Comparison of the contents of common nutrients, chlorophyll and phytol in king grass silage at different cutting heights

Items	Cutting height				SEM
	T1	T2	T3	T4	
Basic nutrient contents					
Moisture/ (g/kg FM)	80.9	83.4	83.2	83.7	0.62
CP/ (g/kg DM)	195^a	178^b	162^c	130^d	0.91
NDF/ (g/kg DM)	444^d	477^c	520^b	591^a	1.01
EE/ (g/kg DM)	28.3^a	27.0^b	25.8^c	21.2^d	0.16
Ash/ (g/kg DM)	126^a	110^b	99^c	77^d	0.58

Chapter 2 Changes of photosynthetic pigments and phytol in herbages

续表

Items	Cutting height				SEM
	T1	T2	T3	T4	
NFC/ (g/kg DM)	209^a	207^a	195^b	183^c	1.24
Chlorophyll and phytol					
Chlorophyll a/ (g/kg DM)	1.87^a	1.42^b	1.14^c	0.72^d	0.77
Chlorophyll b/ (g/kg DM)	0.14^a	0.15^b	0.11^c	0.09^d	0.04
Chlorophyll a+b/ (g/kg DM)	2.01^a	1.57^b	1.25^c	0.51^d	0.35
Phytol/ (g/kg DM)	3.04^a	1.75^b	1.62^c	1.36^d	0.44

Table 2.19 Comparison of the changes of chlorophyll and phytol king grass at different cutting heights before and after ensiling

Items	Cutting height				SEM
	T1	T2	T3	T4	
King grass					
Chlorophyll a/ (mmol/kg DM)	7.11^a	5.00^b	4.12^c	2.59^d	0.51
Chlorophyll b/ (mmol/kg DM)	2.36^a	1.47^b	1.04^c	0.81^d	0.29
Chlorophyll a+b/ (mmol/kg DM)	9.47^a	6.47^b	5.16^c	3.39^d	0.42
Phytol/ (mmol/kg DM)	10.54^a	6.42^b	5.14^c	3.99^d	0.17
King grass silage					
Chlorophyll a/ (mmol/kg DM)	2.09^a	1.59^b	1.28^c	0.81^d	0.82
Chlorophyll b/ (mmol/kg DM)	0.15^a	0.17^a	0.12^b	0.10^b	0.09
Chlorophyll a+b/ (mmol/kg DM)	2.25^a	1.76^b	1.40^c	0.91^d	0.42
Phytol/ (mmol/kg DM)	10.27^a	5.91^b	5.47^c	4.59^d	0.51
King grass/King grass silage					
Chlorophyll a	0.29	0.32	0.31	0.31	0.09
Chlorophyll b	0.07	0.11	0.12	0.13	0.06

续表

Items	Cutting height				SEM
	T1	T2	T3	T4	
Chlorophyll a+b	0.24	0.27	0.27	0.27	0.08
Phytol	0.97	0.92	1.07	1.15	0.04

Table 2.20 Fermentation indexes of king grass silage at different cutting heights

Fermentation quality	T1	T2	T3	T4	SEM
pH	3.91[a]	3.86[b]	3.62[c]	3.95[a]	0.22
Lactic acid/ (g/kg DM)	2.04[b]	1.93[b]	3.46[a]	2.07[b]	0.19
Acetic acid/ (g/kg DM)	8.26[a]	6.70[b]	0.61[d]	4.31[c]	0.48
Propanoic acid/ (g/kg DM)	0.15[a]	0.21[a]	0.02[b]	0.12[a]	0.02
Butyrate/ (g/kg DM)	0.41[a]	0.41[a]	0.26[b]	0.23[b]	0.03
VBN/%	5.84[a]	5.44[c]	5.67[b]	5.94[a]	0.64

Chapter 3　Phytanic aicd production in the rumen

3.1　Effects of fertilization levels and harvesting stages of fresh herbages on ruminal phytanic acid production *in vitro*

3.1.1 Introduction

In above Chapter, I observed the change of photosynthetic pigment and phytol in herbage, and found that the phytol content was affected by harvesting stages and fertilization levels. It was shown in many reports that VFA production and microorganism concentration in the rumen were affected by the contents of NDF and CP in diets (Haaland et al., 1982; Suzuki et al., 2014). Thus, phytanic acid production in the rumen may be affected by the dietary components. There were no reports on the change of phytol and phytanic acid in the rumen, so that the effects of herbages planted in different conditions on phytanic acid production in the rumen is unclear.

Therefore, the study aimed to investigate the effects of N fertilization levels and harvesting stages of fresh herbages on ruminal phytanic acid production.

3.1.2 Materials and methods

3.1.2.1 Incubation herbages

The same IR harvested in Experiment 3 was used in this experiment. The seeds were sown in October 2013. Three fertilization levels were applied at the rate of 0 kg N/ha (Control), 60 kg N/ha (60N) and 120 kg N/ha (120N). The IR was harvested at the booting (April 25th, 2014) and heading (May 9th, 2014) stage. The fresh herbage obtained from one site (3 N fertilization × 2 harvesting stages, total 6 samples) were used. The harvested fresh herbages was immediately preserved in an ice box, and cryopreserved at -30 ℃ within 3 hours after harvesting. Chemical contents of the incubated samples were shown in Table 3.1.

3.1.2.2 Incubation experiment

All animal procedures were managed according to the guidelines of the Animal Care and Use Committee of Hiroshima University. The samples harvested at different harvest stages and different fertilization levels were freeze-dried and ground, and used as substrates for incubation. Two adult wethers (average initial body weight, 48.2 kg) fitted with a rumen cannula were used as donors of ruminal fluid. The wethers were fed basal diets of 50% oats (L) hay and 50% soyabean meal plus barley at maintenance energy level, and were free access to clean drinking water. Rumen fluid was collected through the rumen cannula at 2 h after feeding and diverted to plastic bottles. These fluids were filtered through four layers of cheesecloth and then combined on an equal volume basis. The combined filtrate was mixed with CO_2-bubbled McDougal's artificial saliva (McDougal 1948) at a ratio of 1 : 2 (v/v). Then 30 mL buffered rumen fluid was transferred to 50 mL serum bottles containing 0.3 g sample, and flushed with CO_2 gas. The bottles were capped with a butyl rubber stopper and sealed with an aluminum cap. Incubations were performed in triplicate at 39 ℃ for 48 h in a water bath with a reciprocal shaker (50 strokes/min).

Chapter 3 Phytanic aicd production in the rumen

Triplicate bottles containing only the buffered rumen fluid were also incubated at the same time as a blank. After incubation, the bottle was put into cold water, then the bottle was unsealed and pH was measured after determining gas production volume by a syringe. One mL of incubated fluid was collected and stored at -30 ℃ for VFA analysis, then remained whole content were preserved at -30 ℃ for later analysis.

3.1.2.3 Chemical analysis

The ground dried samples of fresh herbage was analyzed to determine the contents of DM, CP, crude ash and EE according to AOAC (1999), and NDFom according to Van Soest et al. (1991). The NFC content was calculated as follows: NFC = 100 - CP - EE - NDF - crude ash (NRC 2001).

The freeze-dried incubated samples (30 mg) were extracted three times with 80% acetone (Porra et al., 1989), yielding a final volume of 30 mL extract. Then, 1-nonadecanol (0.25 mg/mL hexane) were added (1 mL) to the extract as an internal standard for the phytol analysis. The analysis methods of total and free phytols, pH and VFA were the same as described in Experiment 2.

Phytanic acid in the freeze-dried incubated samples was analyzed according to the methyl esterification method for fatty acid analyses of feeds (Sukhija and Palmquist 1988). Tridecanoic acid (0.25 mg/mL, 1 mL) was used as an internal standard solution. In a screw-capped tube, the internal standard solution (1 mL) was added to 0.1 g freeze dried sample, then dried under an N_2 steam at 40 ℃. Then, 3 mL 0.78 N hydrochloric (HCl) in methyl alcohol and 2 mL chloroform were added to the tube, and the tube was heated for 2.5 h at 65 ℃. Then 5 mL solution of 6% K_2CO_3 and 3 mL hexane were mixed completely. After centrifugation, the supernatant was loaded onto a column containing 0.5 g florisil, and then the column was eluated with 5 mL hexane with diethyl ether (95:5 V/V). The eluate was dried at 40℃ under a constant stream of N_2 gas and redissolved in 1 mL hexane for gas chromatography mass spectrometer (QP2010, Ultra, Shimadzu, Kyoto, Japan) equipped with SP-2560 (100 m × 0.25 mm, film thickness 0.2 μm, supelco). Helium was used as a carrier gas. The column pressure was set at 170 kPa. The initial temperature of the column oven at 70 ℃ for 3 min was raised to 130 ℃ by 11 ℃/min, then to 160 ℃ by 1 ℃/min, finally raised to 220 ℃ by 3 ℃/min. The split ratio was 60.0. The injection volume was 1 μL. For the mass spectrometer, ion source temperature and interface temperature were set at 200 ℃ and 240 ℃ respectively. Selected ion mode was used to measure the relative intensity of 101 and 87 m/z fragmats as target ions of methyl ester of phytanic acid and nonadecanoic acid, respectively.

3.1.2.4 Calculation and statistical analysis

Calculation method:

Ratio (%) of phytanic acid production (PA) = $A/B \times 100$

Ratio (%) of remained phytol (RP) = $C/B \times 100$

Ratio (%) of disappeared phytol (DP) = $100-PA-RP$

A: The mol of phytanic acid produced from 1g sample during incubation

B: The mol of total phytol in 1g sample

C: The mol of total phytol remained after incubation of 1 g sample

Phytanic acid production (mg/g) =

Total phytanic acid after incubation (mg) − blank phytanic acid (mg)

Sample weight (g) used for incubation

Total phytol remained after incubation (mg) − blank phytol (mg)

Sample weight (g) used for incubation

Statistical analysis was performed using the GLM procedure of Statistical Analysis System (SAS 2004). The data of the gas production, pH, VFA, phytanic acid and phytol were subjected to two-way analysis of variance. Tukey's test was used to identify the differences ($P<0.05$) between the treatments.

3.1.3 Results

After 48h incubation, gas production decreased ($P<0.001$) with increasing N fertilization levels (Table 3.2). The pH value increased ($P<0.001$) with increasing N fertilization levels and was lower ($P<0.05$) for the booing stage than that for the heading stage. Total VFA concentration decreased ($P<0.01$) with increasing N fertilization levels, while it was lower ($P<0.001$) for the heading stage compared with the booting stage. Acetic acid and valeric acid productions were affected ($P<0.001$) by both N fertilization and harvesting stage. Phytanic acid production increased ($P<0.001$) with increasing N fertilization levels, and was higher ($P<0.01$) for the booting stage than that for the heading stage (Figure 3.1). The interaction effect for the phytanic acid production was also significant ($P<0.01$). The phytanic acid production for the 60N and 120N treatments at the heading stage did not differ. The ratio of phytanic acid production (12% ~ 17%) decreased ($P<0.01$) with increasing N fertilization levels, and was higher for the booting stage than that for the heading stage ($P<0.01$). In relative term, most of phytol in the herbage (66%~74%) was remained in the incubation medium. In addition, a small part (12%~23%) of phytol in the herbage was disappeared (non-remain phytol and non-phytanic acid) during incubation.

3.1.4 Discussion

After 48 h incubation, decreased gas production with increasing N fertilization levels probably reflected higher fiber and lower NFC contents in fresh herbage at the higher fertilization levels. The changes of pH and total VFA concentration were consisted with the changes of gas production. The differences in the acetic acid and valeric acid production with harvesting stage were probably due to the difference of fiber content in the incubated herbages.

Phytanic acid production was affected by the N fertilization levels and harvesting stages, and these responses were probably determined by the phytol content in the incubated herbages. No differences were found in phytanic acid production between 60N and 120N treatments for heading stage. This results was owing to the effects of chemical component in the incubated samples, because the contents of CP and NDF in the incubated samples had little difference. Phytanic acid production ratio was also significantly affected by N fertilization levels and harvesting stages and their interaction. However, the differences between treatments were small and ranged from 12% to 17%. The reason for the such interactive effect was not clear in this experiment. Therefore, although the amount of phytanic acid production increased with increasing phytol content of herbage, its production ratio decreased. The factors affecting phytanic acid production related to rumen fermentation needs to be verified in future research.

After 48 hours incubation, about 66%~74% of phytol was remained in the incubation medium. These residual phytol seems to be a esterified form, because free phytol was not detected. As described in Chapter 2, phytol in silage also existed in an esterified form. This may be related to the chemical propety of phytol. Thus, the phytol existed in the rumen as an esterified form, which might be associated with some fatty acids or other compounds. In addition, about 12%~23% phytol disappeared (unaccountable) after incubation, which might be converted to other substances, such as phytene (Ueda et al., 2016).

In conclusion, the phytanic acid production was higher for fresh herbages at higher N fertilization levels and at booting stage. The ratio of ruminal phytanic acid production was 12%~17%, and most of phytol (66%~74%) was remained and existed as an esterified form.

Chapter 3 Phytanic aicd production in the rumen

Table 3.1 Chemical composition in dried samples of fresh herbages used for in vitro incubation

Item	Nitrogen fertilizer application rate					
	Cotrnol		60N		120N	
	Booting	Heading	Booting	Heading	Booting	Heading
Moisture (%)	76.7	69.3	82.5	84.2	86.3	81.4
Crude ash (DM %)	5.86	5.61	7.35	7.56	9.02	7.66
Crude protein (DM %)	6.25	5.13	8.41	8.27	16.25	10.66
Ether extract (DM %)	1.91	1.58	2.48	2.27	3.40	2.51
NDF (DM %)	34.5	41.2	40.6	51.6	46.4	51.1
NFC (DM %)	51.5	46.5	41.2	30.3	24.9	28.1
Phytol (mg/g)	0.91	0.84	1.61	1.69	1.77	1.88

NDFom, neutral detergent fiber exclusive of residual ash; NFC, non-fibrous carbohydrate.
Fresh herbages were applied nitrogen fertilization at 0 kg N/ha (Control), 60 kg N/ha (60N) and 120 kg N/ha (120N), and harvested at the booting stage and the heading stage.

Table 3.2 In vitro fermentation parameter after 48h incubation of dried sample of fresh Italian ryegrass herbage grown under different fertilization levels (control, 0 kg N/ha; 60N, 60 kg N/ha; 120N, 120 kg N/ha) and harvested at booting and heading stages

Item	Nitrogen fertilizer application rate						SEM	P-value		
	Cotrnol		60N		120N					
	Booting	Heading	Booting	Heading	Booting	Heading		N	H	N*H
Gas production (ml/g DM)	50.7	52.0	48.7a	45.3b	48.0	46.7	0.694	<0.001	0.068	0.043
pH	5.91	5.94	5.97b	6.12a	6.06	6.08	0.025	<0.001	0.010	0.041
TVFA (mmol/L)	105	103	103	97	104	97	1.183	0.009	<0.001	0.190
Acetic acid (mol %)	44.8	44.9	44.7	44.7	43.9	44.1	0.091	<0.001	0.021	0.086
Propionic acid (mol %)	39.6	39.3	39.2	38.6	39.7	38.5	0.120	0.100	0.080	0.090
Butyric acid (mol %)	11.3	11.4	11.4	11.5	11.4	12.1	0.126	0.078	0.048	0.062
Valeric acid (mol %)	1.88	1.91	1.98b	2.21a	2.20b	2.31a	0.012	<0.001	<0.001	<0.001
A/P	1.12	1.14	1.14	1.16	1.11	1.15	0.005	0.055	0.091	0.088

a,b Means within same fertilizer application rate marked with different superscripts differ significantly ($P<0.05$)

TVFA, total volatile fatty acid; A/P, the ratio of acetic acid to propionic acid; SEM, standard error of the mean; N, effect of nitrogen fertilizer application rate; H, effect of harvest stage; N*H, interaction effect between N and H.

Chapter 3 Phytanic aicd production in the rumen

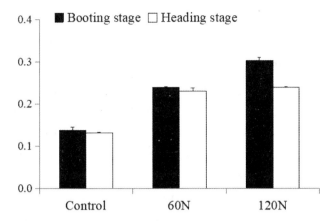

Figure 3.1 Phytanic acid production after 48 hours in vitro incubation

The value of phytanic acid production means phytanic acid (mg) produced from 1 g incubated sample during 48 hours.

a, b Means within heading stage marked with different superscripts differ significantly ($P<0.05$)

x, y, z Means within heading stage marked with different superscripts differ significantly ($P<0.05$)

N, effect of nitrogen fertilizer application rate; H, effect of harvest stage; N*H, interaction effect between N and H. **, $P<0.01$.

3.2 Effects of fertilization levels and harvesting stages of Italian ryegrass silages on ruminal phytanic acid production *in vitro*

3.2.1 Introduction

The results of Experiment 5 showed that the ratio of ruminal phytanic acid production from phytol in the fresh herbage sample was only 12%~17%, and a part of phytol was disappeared during in vitro incubation. At the same time, most remained phytol existed as an esterified form. As indicated in Chapter 2, although chlorophyll greatly decreased during ensiling, phytol content did not change and the phytol moiety from decomposed chlorophyll was presumably existed as an esterified form. Therefore, extent of phytanic acid production from silages by rumen incubation may be different from those produced from fresh herbage, if the re-esterified phytol moiety is converted to phytanic acid more easily. These properties of silage phytol may modify the effect of N fertilization levels and harvest stages on phytanic acid production from herbages.

Grass herbage is rich in unsaturated fatty acids. Some unsaturated fatty acid such as ω-3 fatty acids in herbage and cis-9 trans-11 conjugated linoleic acid (CLA) produced in the rumen had positive effects on human health. CLA has the functions to prevent cancer and to improve immune response (Belury, 2002; Palmquist et al., 2005). The beneficial ω-3 unsaturated fatty acid would decrease rapidly after the bio-hydrogenation of fatty acid in the rumen. Therefore, exploring the changes of fatty acid in the rumen is also meaningful.

Therefore, the objective of this experiment was to investigate the effects of N fertilization levels and harvesting stages of silages on ruminal phytanic acid production and fatty acid composition in vitro (Experiment 6). In addition, the change of phytanic acid production in the rumen was unclear. Therefore, the changes of phytanic acid production during incubation period were explored (Experiment 7).

3.2.2 Materials and methods

3.2.2.1 Incubation samples

The same silage samples prepared in Experiment 3 were used in this experiment as incubation substrate. The herbages at 0 kg N/ha (Low-N) and 120 kg N/ha (High-N) were harvested at booting (April 25th, 2014) and heading (May 9th, 2014) stage respectively. Freeze-dried silage samples of Low-N (0 kg N/ha) and High-N (120 kg N/ha) treatments, which were harvested from the same site (one of three sites) at the booting and heading stages were used. Harvesting site

Chapter 3 Phytanic aicd production in the rumen

for the silage used in this study was same as Experiment 5, and the ensiling methods were same as Experiment 2. The freeze-dry samples of silage were preserved at -30 ℃ till the incubation. The chemical compositions of samples were shown in Table 3.3.

3.2.2.2 Incubation procedures

All animal procedures were managed according to the guidelines of the Animal Care and Use Committee of Hiroshima University.

Experiment 6

The freeze-dried silage samples for the different harvesting stages and different fertilization levels were used as substrates for incubation. The donner weathers, diet for the animal and all feeding methods were same as Experiment 5. Rumen fluid was collected from two wethers through the rumen cannula at 2 h after feeding and diverted to plastic bottles, then the filtered rumen fluid and the buffer were mixed as incubation fluid [combined filtrate : artificial saliva = 1 : 2 (v/v)]. Ground sample (0.3g) was put into the serum bottle, and incubated for 48 hours by the same methods described in Experiment 5.

Experiment 7

To investigate the time course change of phytanic acid production, the silage harvested at booting stage at 120 kg N/ha was used as samples, and incubated for 3, 6, 9, 12, 18, 24, 36 and 48 hours by the same method described in Experiment 5.

3.2.2.3 Chemical analysis

The analysis methods for chemical components, phytol, phytanic acid, pH and VFA were same as Experiment 5. Fatty acid content in incubation residue were also measured by the same analytical methods of phytanic acid. The mixture of fatty acid methyl ester (18920, Supeclo, Sigma-Aldrich, Tokyo, Japan) mixed with methyl trans-11 vaccinate (V1381, Sigma-Aldrich, Tokyo, Japan), methyl cis-9 trans-11 octadecanoate (1254, Matreya, Kanto chemical, Tokyo, Japan) and phytanic acid methyl ester (P3819, Sigma Aldrich, Tokyo, Japan) was used as a standard solution.

3.2.2.4 Statistical analysis

Statistical analysis was performed using the GLM procedure of Statistical Analysis System (SAS 2004). The data of the gas production, pH, VFA, phytanic acid, phytol and fatty acid in Experiment 6 were subjected to two-way analysis of variance. The data for the time course change in Experiment 7 was analysed by one-way ANOVA. Tukey's test was used to identify the differences ($P<0.05$) between the treatment or time. The calculation methods for PA, RP and DP were same as Experiment 5.

3.2.3 Results

After 48 hours incubation, the gas production was higher ($P<0.001$) for the Low-N compared with the High-N (Table 3.4). For Low-N, gas production was higher at the booting stage, while the opposite trend was observed at the heading stage. The pH was not affected by the harvesting stage, however, it was higher ($P<0.001$) for the High-N compared with the Low-N. TVFA concentration was not affected by neither N fertilization level nor harvesting stages. However, the molar proportions of VFA except for acetic acid were affected ($P<0.05$) by the N fertilization levels (Table 3.4). The C12:0, C14:0, C16:0, 11 C18:1, -9 C18:1, C18:3 and phytanic acid remained after incubation were higher for the High-N treatment compared with the Low-N (Table 3.5). Phytol content in the incubation residue was also higher for the High-N treatment (Table 3.5). The harvesting stage did not affect fatty acids remained after incubation. In Experiment 7, amount of phytol in the incubation residue decreased ($P<0.05$) rapidly at the first 3 hour incubation (Figure 3.2). However, the phytanic acid production increased gradually. After 48 hour incubation, about 15% phytol in herbage was converted to phytanic acid (Figure 3.2).

Chapter 3　Phytanic aicd production in the rumen

3.2.4 Discussion

After 48 hour incubation, phytanic acid production was not affected by harvesting stages, which reflected the small differences in phytol content in the silage between the booting and heading stage. In addition, phytanic acid production processes may be affected by the rumen fermentation. The relatively higher ratio of phytanic acid production at the Low-N fertilization levels was consisted with the results obtained from the dried fresh herbage (Experiment 5). After incubation, about 40%~60% dietary phytol was remained, and about 12%~24% dietary phytol was unaccounted. Phytanic acid production ratio for the silage (17%~38%) was higher than that for the fresh herbages (12%~17%) obtained in Experiment 5. As described in Chapter 2, part of phytol in silage was once released from chlorophyll, which might be easily converted to phytanic acid. In this experiment, most parts of dietary phytol were remained and a small part was disappeared after incubation, which was the same as the results of Experiment 5.

After incubation, most of the unsaturated fatty acids were saturated owing to the bio-hydrogenation in the rumen. Based on the fatty acid composition, unsaturated fatty acids accounted for 20% of the total fatty acid remained after incubation. In addition, C18 fatty acids accounted for 81% of total fatty acids in silage. Although unsaturated fatty acids accounted for about 99% of total C18 in silage, after 48 hour incubation, the ratio of unsaturated fatty acid in total C18 reduced to 21% and 45% for the Low-N and High-N treatments, respectively. N fertilization could contribute to reduce bio-hydrogenation of unsaturated fatty acid in the rumen.

In Experiment 7, the production ratio of phytanic acid was about 15% after 48 hours incubation. However, the initial dietary phytol was remained only 20% and most parts of phytol was disappeared, in which this results was different from Experiment 5. Although the reason for this discrepancy was unclear, these results might be affected by rumen fermentation condition. At the first 3 hours incubation, the phytol decreased rapidly, while the production rate of phytanic acid was slow. From 3 to 48 hour incubation, both the rates of phytol disappearance and phytanic acid production were low. These results verified that the production rate of phytanic acid was slow in the rumen incubation. In addition, the ratio of the disappeared phytol might be related to the nature of incubated rumen fluid.

In conclusions, the phytanic acid production was higher for silage at higher N fertilization levels and at booting stage. The ratio of ruminal phytanic acid production from phytol was 17%~38%, and most of phytol (40%~60%) existed as an esterified form. The ratio of phytanic acid production to total phytol in herbages was higher for silage (15%~36%) compared with those for fresh herbages (12%~17%).

Table 3.3 Chemical composition and fatty acid in dried samples of silage used for in vitro incubation

Item	Nitrogen fertilizer application rate				
	Low-N			High-N	
	Booting	Heading		Booting	Heading
Chemical composition					
Moisture (%)	63.8	61.3		75.6	62.4
Crude protein (DM %)	8.1	6.1		14.8	10.7
Ether extract (DM %)	2.2	2.2		4.1	3.0
NDFom (DM %)	38.1	53.3		55.1	56.7
Crude ash (DM %)	7.3	7.8		11.7	8.8
NFC (DM %)	44.3	29.8		14.4	20.8
Phytol (mg/g)	0.83	0.94		3.32	2.11
Fatty acid (mg/g DM)					
C12:0	0.021	0.021		0.025	0.022

Chapter 3 Phytanic aicd production in the rumen

C14:0	0.062	0.058	0	0.085	0.064
C15:0	0.007	0.017	0	0.024	0.025
C16:0	2.327	2.228	4	4.188	2.969
C17:0	0.015	0.016		0.023	0.020
C18:0	0.169	0.161		0.236	0.184
cis-9 C18:1	0.157	0.185		0.104	0.260
cis-9, 12 C18:2	1.766	1.861		3.234	1.973
cis-9, 12, 15 C18:3	8.698	8.156		18.705	9.535
cis-9 trans-11 CLA	0.143	0.065		0.172	0.048
C22:0	0.175	0.009		0.234	0.192

NDFom, neutral detergent fiber exclusive of residual ash; NFC, non-fibrous carbohydrate.
Silage were applied nitrogen fertilization at 0 kg N/ha (Low-N) and 120 kg N/ha (High-N), and harvested at the booting stage and the heading stage.

Table 3.4. In vitro Fermentation parameter after 48h incubation of different fertilization levels (Low–N, 0 kg N/ha; High–N, 120 kg N/ha) and harvest stage (booting and heading stages) Italian ryegrass silage

Item	Nitrogen fertilizer application rate				SEM	P-value		
	Low-N		High-N			N	H	N*H
	Booting	Heading	Booting	Heading				
Gas production (ml/g DM)	48.7	45.3	36.7	40.0	0.54	<0.001	0.122	0.004
pH	6.13	6.21	6.35	6.35	0.03	<0.001	0.161	0.179
Total VFA (mmol/L)	87.9	80.9	81.9	77.1	3.31	0.179	0.110	0.745
Acetic acid (mol %)	45.0	45.4	45.1	45.3	0.29	0.921	0.286	0.835
Propionic acid (mol %)	40.0	39.9	35.4	36.6	0.93	0.003	0.568	0.519
Butyric acid (mol %)	10.4	10.1	12.8	11.8	0.86	0.043	0.471	0.689
Isovaleric acid (mol %)	1.89	1.93	2.74	2.59	0.13	0.001	0.657	0.498

续表

Item	Nitrogen fertilizer application rate				SEM	P-value		
	Low-N		High-N			N	H	N*H
	Booting	Heading	Booting	Heading				
Valeric acid (mol %)	1.78	1.69	2.50	2.32	0.13	0.001	0.334	0.716
A/P	1.12	1.14	1.27	1.24	0.03	0.002	0.768	0.005

Means within same fertilizer application rate marked with different superscripts differ significantly (P<0.05).

VFA, volatile fatty acid; A/P, the ratio of acetic acid to propionic acid; SEM, standard error of the mean; N, effect of nitrogen fertilizer application rate; H, effect of harvest stage; N*H, interaction effect between N and H.

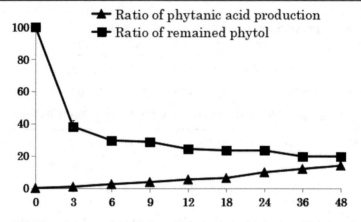

Figure 3.2 The change of phytol and phytanic acid during 48 h incubation

The ratio of phytanic acid production means the relative ratio of phytanic acid production to initial phytol in silage.

The ratio of remained phytol means the relative ratio of remained phytol to initial phytol in silage.

Chapter 4 Effect of forages sources on phytanic acid content in milk of dairy cows

Chapter 4 Effect of forages sources on phytanic acid content in milk of dairy cows

4.1 Introduction

It was shown in Chapter 2 that chlorophyll was decomposed gradually, but the phytol was well preserved during ensiling. In addition, a previous study showed that the phytanic acid content in milk was higher in cows fed silage compared with those fed hay (Markus et al., 2012). These suggested that the phytanic acid content in milk would reflect dietary phytol content.

The IR silage as well as whole crop corn silage are used extensively for dairy cow feeding as major forage sources. The forage sources (IR vs corn silage in diets) may affect the milk production as well as the content of milk components such as fatty acids and phytanic acids. The differences in fatty acid profile could affect the value of milk, because the fatty acid profile in milk may affect human health. Whole crop corn silage contains leaves, steams, grains and cobs, so that chlorophyll or phytol content would be diluted with non-leaf part of plant. Thus, the phytol content is expected to be higher in IR silage compared with corn silage. In dairy production systems, total mixed rations (TMR) containing forages, grains, protein feeds, minerals, vitamins and feed additives are used to satisfy the nutrient requirement of cows (Chen et al., 2015; Nishino et al., 2003). The phytanic acid content in milk of cows fed TMR containing IR silage is expected to be higher than those of cows fed TMR containing corn silage, due to the difference in the phytol content between IR silage and corn silage. At present, although some papers reported the phytanic acid content in milk products (Markus et al., 2011; Vetter and Schroder, 2010), there were no reports confirming the relationship between the phytol intake and phytanic acid in milk. Therefore, this study aimed to explore the effect of forage sources (IR silage vs corn silage) in TMR on milk production, milk composition and phytanic acid content in milk, and the extent of conversion of dietary phytol to milk phytanic acid.

4.2 Materials and methods

4.2.1 Experimental design and animals

All animal procedures were managed according to the guidelines of the Animal Care and Use Committee of Hiroshima University. Total 17 Holstein cows (8 primiparous and 9 multiparous cows) averaging 2 parity, 213 days in milk (DIM) and 732 kg of body weight (BW) were used in the experiment consisting three 21 days periods at Hiroshima University Farm. Cows were raised in the cowshed installing an automatic milking system (Astronaut A3 next, Lely, the Netherland) and the roughage intake control system (Insentec, Drachten, the Netherland). Cows were supplied a concentrate diet with an automatic feeder in the automatic milking system. The corn silage TMR were fed during the first and third period (period 1 and period 3), and the IR silage TMR were fed during the second period (period 2). The ingredients and chemical composition of the TMR were shown in Table 4.1. The milk samples were collected at the last 2 days of each period, and preserved at -30 ℃ for the later analysis after determination of the milk component contents. The blood of caudal artery was collected at 13:00 on the last day of each period. Then, plasma was collected after centrifugation and preserved at -30 °C for further analyses.

Feed samples were collected for last 3 days and freeze dried for later analysis. Spot feces samples were collected from the 4 cows during final 3 days of each period. The feces were collected immediately after defecation at morning, afternoon and evening, and composited for each cow. Then, the fecal samples were freeze dried for later analysis.

4.2.2 Chemical analysis

Feed samples were analyzed for DM, crude ash, CP and EE by the methods of AOAC (1990). Milk samples were measured for fat, protein, lactose and solids-not-fat (SNF) by an infrared analyzer (Lactoscope Filter C4+, Delta Instruments, Drachten, and the Netherlands). Plasma samples were analyzed for glucose, non-esterified fatty acid (NEFA), triglyceride (TG), total cholesterol (T-CHO), blood urea nitrogen (BUN) using an automated biochemical analyzer (AU 480; Beckman Coulter Brea, CA, USA). The fatty acid content including phytanic acid in milk samples were determined by GC-MS after acid methylation of lipid extracts as follows. First, 1 mL of milk was mixed with 0.2 mL 28% ammonia solution and 0.8 mL 96% ethanol. Then, 0.2 mL of methyl tridecanoate (2.5 mg/ml hexane) was added as an internal standard. 1 mL 0.025% BHT diethyl ether and 1 mL hexane were added and mixed well. The supernatant after centrifugation at 1710 for 5 min was corrected. This extraction was repeated and combined supernatant was dried under an N_2 steam. The dried extract was methyl esterified by the methods as described in Experiment 5. The analytical conditions of GC-MS for fatty acids and phytanic acid measurement were same as Experiment 6. Phytol in TMR and feces, and phytanic acid in feces were also measured by the methods described above.

4.2.3 Statistical analysis

Statistical analysis was performed using the GLM procedure of

Statistical Analysis System (SAS 2004). The data were analyzed as a complete blocked design. Tukey's test was used to identify the differences of means ($P<0.05$) among experimental periods.

4.3 Results

The chemical compositions of the two TMR were shown in Table 4.1. Chemical compositions of the TMR had no differences in DM, CP and NDF among the periods. The phytol content in the TMR at period 2 (IR silage TMR) was higher ($P<0.05$) than that at period 1 and 3 (corn silage TMR). DMI of TMR and concentrate for period 1, 2 and 3 were 19.5, 19.7, 20.1 kg/d and 4.9, 4.8, 4.6 kg/d, respectively, and there were no differences among the periods (Table 4.2). The average milk yields were about 28.4 kg/day which was similar among the periods. There were no differences in fat, protein, lactose, SNF, SCC content in milk between the periods (Table 4.2). There were no differences in fatty acid composition in milk among the periods (Table 4.3). There were no differences in the plasma concentration of glucose, TG, T-CHO, NEFA among the periods, while the BUN was higher ($P<0.05$) at the period 2 (Table 4.4). The content and secretion of milk phytanic acid were higher ($P<0.05$) at the period 2 compared with the period 1 and 3 (Table 4.5). In addition, phytanic acid content in feces was also higher ($P<0.05$) at the period 2. Phytol contents in feces had no differences among the three feeding periods (Table 4.5).

4.4 Discussion

This experiment aimed to explore the conversion ratio of dietary

Chapter 4 Effect of forages sources on phytanic acid content in milk of dairy cows

phytol to milk phytanic acid in cows fed TMR having different phytol contents. No differences in DMI, milk yields and milk component among the periods indicates that the difference of silage in TMR did not affect milk production performance due to the similar energy intake. The result also consistent with other reports (O'Mara et al., 1998; Larsen et al., 2012; Rezaei et al., 2015; Paiva et al., 2016). Phytanic acid content in milk was higher for cows fed the IR silage TMR. Markus et al. (2011) reported that phytanic acid content in milk was between 0.021 and 0.2 mg/g milk. However, the phytanic acid content in this experiment was lower than that of their report, presumably due to the low phytol content in the TMR used in this study. In this experiment, silage and hay (Oats hay and alfafa hay) accounted for 50% of TMR, while Markus et al. (2011, 2012) used the diet containing 86% of silage and hay. In addition, the feeding conditions and diets of cows were also important effective factors for milk quality (Dahl,1997; Blanch,2016).

The phytol intakes from TMR were calculated to be 9.5, 15.5 and 10.4 g/d for period 1, 2 and 3 respectively. The phytol intake during period 2 was obviously higher ($P<0.05$) compared with other periods. In addition, the total phytanic acid secretion into the milk was calculated to be 0.27, 0.42 and 0.25 g/d for period 1, 2 and 3 respectively, Based on these calculation, the conversion ratio of dietary phytol to milk phytanic acid was estimated to be only 2.6%.

The feces of 4 cows were collected at the final stage of the experiment. Phytanic acid was found in the feces, and the contents were 1.2, 1.6 and 1.2 mg/kg for period 1, 2 and 3, respectively. These results indicate that not all the phytanic acid produced in the rumen could be absorbed in cows, and part of it was excreted into the feces. Slightly higher phytanic acid content in the feces was observed for the period 2. This higher excretion was also affected by phytol intake. Because the TDN content of TMR diets was about 70% for the three periods, DM digestibility of the TMR can be assumed to be 70%. Based on this assumption, fecal excretion of phytanic acid was estimated to be very small, only 0.07% of phytol intake. Thus, most of dietary phytol was not recovered as phytanic acid in milk as well as in feces. This low appearance of phytanic acid was presumably owing to the low phytanic acid production in the rumen found in Chapter 3. Therefore, most phytanic acid in the rumen may be utilized other than milk component. In addition, phytol was found in feces, and there were no differences in the phytol content among the three periods. Using above assumption. The fecal excretion of phytol was estimated to be 3.3, 2.8 and 2.9 g/d for period 1, 2 and 3 respectively. This phytol excretion accounted for 35% (period-1), 19% (period-2) and 30% (period-3) of dietary phytol, respectively. Although the phytol intake of cows fed IR silage TMR was higher, the excreted ratio was lower for the cows fed IR. Compared with corn silage TMR, the apparent use of phytol in the total digestive tract seems to be higher for IR silage TMR.

In the experiment, there were no differences in fatty acid profile in milk. The results consisted with other reports (Razzaghi et al., 2015). Herbage is usually rich in C18:3 fatty acid (Dierking et al., 2010), so that C18:3 in milk is one of the important fatty acid marker in some organic milk systems. The relationship between the C18:3 in milk fat and phytanic acid production was observed in some reports, and showed that C18:3 were 3 times higher than phytanic acid (Markus et al., 2011). However, no differences in C18:3 were found among the three feeding periods in this experiment. Although milk C18:3 could reflect the fatty acid composition in diets, it could not be regarded as a marker for phytanic acid content in milk.

The concentration of BUN in the period 2 was higher than that in period 1 and 3. The different BUN concentrations among the period might be due to a balance between hepatic production and output (urinary excretion and recycling) of urea-N (Radostits et al., 2007). BUN is affected by protein and energy consumed by animal and breakdown of muscle protein (Reist et al., 2003). In addition, it was reported that there was a positive correlation between BUN and ruminal ammonia (Huyen et al., 2012; West et al., 1993). Ruminal ammonia

Chapter 4 Effect of forages sources on phytanic acid content in milk of dairy cows

was utilized by rumen microorganism (Cherdthong et al., 2014). Different components of diets or protein would affect ruminal ammonia content (Ouellet and Chiqueete, 2016). In this experiment, thus, the IR silage TMR would have higher degradable N compared with the corn silage TMR.

In conclusion, phytanic acid content in milk was higher for cows fed the IR silage TMR compared with the corn silage TMR. However, conversion ratio of dietary phytol to milk phytanic acid was estimated to be only 2.6%. There were no differences in milk yield and milk composition contents between cows fed the IR silage TMR and corn silage TMR.

Table 4.1 Ingredients and composition of total mixed rations for cows at each experimental period

Item	Period-1	Period-2	Period-3	SEM
Ingredient (% of DM)				
Italian ryegrass silage	0.0	20.4	0.0	
Corn silage	30.2	9.5	30.2	
Oats hay	9.9	7.1	9.9	
Alfafa hay	11.1	12.7	11.1	
Beet pulp	6.6	6.9	6.6	
Concentrate diets	40.2	41.0	40.2	
$CaCO_3$	0.8	1.0	0.8	
Vitamin	0.9	1.0	0.9	
NaCl	0.4	0.4	0.4	
Composition (% of DM)				
Dry matter (% of FM)	44.7	46.0	47.1	5.96
Crude protein	12.6	13.0	13.5	0.56
NDFom	40.3	42.3	41.0	1.66
Ether extract	3.15	3.09	3.10	0.03
Ca	0.85	0.83	0.80	-
P	0.38	0.37	0.36	-
TDN	66.9	68.7	69.3	-
Phytol (g/kg DM)	0.483[b]	0.784[a]	0.517[b]	0.007

[a,b] Means with different letters significantly differ ($P<0.05$).
SEM, standard error of mean; NDFom, neutral detergent fiber exclusive of residual ash; TDN, total digestible nutrients.

Table 4.2 Effects of feeding periods on feed intake, milk yield, and milk composition in dairy cows

Item	Period-1	Period-2	Period-3	SEM
DMI (kg/d)				
TMR	19.5	19.7	20.1	2.33
Concentrate	4.88	4.75	4.60	0.260
Milk yield (kg/d)	28.8	29.2	27.3	1.45
Milk composition (%)				
Fat	4.03	4.14	4.09	0.340
Protein	3.48	3.45	3.56	0.136
Lactose	4.57	4.57	4.54	0.076
SNF	8.97	9.03	9.02	0.183
SCC, ×1,000/ml	138	145	148	12.7

SEM, standard error of mean; DMI, dry matter intake; SNF, solids-not-fat; d, day;
Cows were fed corn silage TMR (period 1 and 3) and Italian ryegrass silage TMR (period 2).

Table 4.3 Effects of feeding periods on fatty acid composition (% of total fatty acid) in milk of dairy cows

Fatty acids	Period-1	Period-2	Period-3	SEM
C8	0.69	0.64	0.89	0.079
C10	3.06	3.05	3.36	0.093
C12	4.60	4.63	4.91	0.073
C14	13.8	13.5	13.9	0.162
C14:1	2.77	2.9	2.87	0.064
C15	1.27	1.32	1.29	0.029
C16	34.7	34.0	34.2	0.411
C16:1	2.32	2.36	2.23	0.057
C17	0.518	0.531	0.503	0.016
C18	5.89	4.61	4.18	0.764
trans-11 C18:1	3.42	6.1	5.72	1.315

续表

Fatty acids	Period-1	Period-2	Period-3	SEM
cis-9 C18:1	24.4	24.5	23.8	0.310
cis-9, 12 C18:2	0.94	0.92	0.96	0.023
cis-9, 12, 15 C18:3	0.59	0.67	0.67	0.035

SEM, standard error of mean.
Cows were fed corn silage TMR (period 1 and 3) and Italian ryegrass silage TMR (period 2).

Table 4.4 Effects of feeding periods on plasma metabolite concentrations in dairy cows

Item	Period-1	Period-2	Period-3	SEM
Glucose (mmol/L)	3.98	3.74	3.99	0.063
TG (μmol/L)	63.9	73.8	70.8	4.14
T-CHO (mmol/L)	5.73	5.81	5.41	0.257
NEFA (μEq/L)	110.1	100.3	112.5	5.10
BUN (mmol/L)	1.96^b	2.72^a	1.98^b	0.109

^{a,b} Means with different letters significantly differ ($P<0.05$).
SEM, standard error of mean; TG, triglyceride; T-GHO, total-cholesterol; NEFA, non-esterified fatty acids; BUN, blood urea nitrogen.
Cows were fed corn silage TMR (period 1 and 3) and Italian ryegrass silage TMR (period 2).

Table 4.5. Effects of feeding periods on milk and feces phytanic acid in dairy cows

Item	Period-1	Period-2	Period-3	SEM
Phytanic acid in milk (mg/kg, n=17)	9.3b	13.9a	8.8b	0.378
Phytanic acid secretion in milk (mg/d, n=17)	269.8b	415.6a	247.5b	14.04
Phytanic acid in feces (mg/kg DM, n=4)	1.15b	1.65a	1.17b	0.022
Phytol in feces (g/kg DM, n=4)	0.515	0.455	0.492	0.035

a,b Means with different letters significantly differ ($P<0.05$).
SEM, standard error of mean; DM, dry matter; d, day.
Cows were fed corn silage TMR (period 1 and 3) and Italian ryegrass silage TMR (period 2).

Chapter 5   General discussion

Since 1990s, consumers pay more attention to the organic or functional ruminant products, such as the functional milk (Bourn and Prescott 2002). Because phytanic acid is considered to have a positive effect on human health, the phytanic acid content in ruminant products was regarded as an important idex to determine the functional property in many studies (Vetter and Schroder 2010). Noteworthy, phytanic acid cannot be synthesized by mammals (Steinberg et al., 1967). Next to ruminant production from forage, the only additional natural source for phytanic acid is fish and other marine organisms (Masters-Thomas et al., 1980; June et al., 1993; Allen et al., 2008). Although the content of phytanic acid in milk was found to be higher for cows fed only grass (Leiber et al., 2005), effective utilization of forage to produce the milk containing higher phytanic acid will be paid more attention. Therefore, effective production of forages containing higher chlorophyll or phytol and effective feeding methods to produce dairy products containing higher phytanic acid are necessary study areas.

This study aimed to explore the changes of phytol in herbage based on the production methods of forages, the factors of phytanic acid production in the rumen, and the secretion of phytanic acid in milk of cows with different diets. Although some unclear subjects were remained, the experimental results provided effective and valuable data for this functional products relating to the feeding management of cows and the preparation and preservation of herbage.

Chapter 5　General discussion

5.1　Factors for the changes of photosynthetic pigments and phytol in herbages

Fertilization is a traditional method to improve crop yields and chemical composition of plants, which are also affected by harvesting stages of forages. In this study, the effects of different fertilization levels and harvesting stages on the content of photosynthetic pigments and phytol in IR were explored. The results showed that early harvest stages and higher fertilization levels could be effective factors to obtain higher photosynthetic pigments and phytol in the herbage.

Using the data of Experiments 1 and 3, the relationship between the CP content and phytol in fresh herbage was tested. The results indicated that the CP content and phytol in fresh herbage was linearly correlated (Figure 5.1). The phytol content could be estimated from the protein content in fresh herbage.

Using the data of Experiments 1 and 3, the changes of chlorophyll was compared when the harvested herbage was dried for 1 day and 7 days. The results showed that there was no difference in the chlorophyll content after the herbage was dried for 1 day (Figure 5.2). The results showed that sufficient moisture content in herbages would be necessary to preserve chlorophyll, at least, for one day.

Moisture of herbage is normally controlled at 60%~70% for the ensilage by the conventional methods (Beaulian et al., 1993). Therefore, the loss of chlorophyll is expected to be little when the moisture was enough, even if the herbage dried under natural conditions for 1 day before ensiling. This also explained that phytol could be preserved well from harvest to ensiling stage. However, the phytol content could decrease after dried for 7 days, so that hay could largely decrease the photosynthetic pigment content. In contrast, the results of Experiments 2, 3 and 4 showed that phytol could be preserved during ensiling. Even if the pH decrease largely by addition of LAB, the preservability of phytol in silage would not be affected.

Using the data of Experiments 3 and 4, CP and phytol content in silage was linearly correlated (Figure 5.3). In Chapter 4, IR silage and corn silage used in Experiment 8 were produced by the roll-bale methods and preserved for more than 5 months. The CP and phytol content in these practical silages were well fitted with the relationship produced by the small scale silages (Figure 5.3). Phytol in roll-ball silage could be also preserved well for more than 5 months under the practical farm condition.

5.2 Factors for the changes of phytanic acid in the rumen

The production of phytanic acid was affected by the action of rumen microorganism to the phytol moiety of chlorophyll (Ackman and Hansen, 1967). Therefore, it could be predicted that the phytanic acid production was affected by phytol content in forage or rumen microbial population. In Chapter 3, the changes of phytanic acid production were explored by incubating the fresh herbage and silage which had different growth conditions. The results confirmed that the fresh herbage or silage with higher phytol content could produce higher phytanic acid in the rumen. Because the contents of CP, NDF and NFC were affected by harvest stages and fertilization levels (Islam and Garcia, 2011; King et al., 2012), the ruminal fermentation of forage was also affected by harvesting stage and fertilization levels of the forages, which could change pH and VFA concentration in the rumen (Suzuki et al., 2014). These changes might affect the phytanic acid production indirectly, through the change of microbial population and activities.

The ratio of phytanic acid production for the silage (Experiment 6) was higher than that for the fresh herbage (Experiment 5). In chapter 2, although the chlorophyll was decomposed largely during ensiling, phytol content did not decrease and existed as an ester form. Therefore, the phytol moiety from decomposed chlorophyll could be combined with other substances during ensiling. The re-esterified phytol might be easily converted to phytanic acid under the action of rumen microorganism during incubation. Some phytol might be derived from the ester of long-chain fatty acid in leaves (Prisker et al., 1989). However, it is unclear whether this phytol could be easily converted to phytanic acid in rumen.

Most of phytol in fresh herbage and silage was remained after the incubation. In Chapter 3, the animals used in the experiment were fed the diet containing 50% concentrate and 50% oats hay which had lower phytol content. These dietary conditions would affect the species and concentration of microorganism in rumen fluid used for the incubation (Moya et al., 2009; Belanche et al., 2012). Further studies on the factors affecting phytanic acid production in the rumen will be necessary.

In summary, it was confirmed that the feeding forage with higher phytol content could increase ruminal phytanic acid production. In addition, ensiled herbage compared with fresh herbage could produce greater amount of phytanic acid. However, the rate of phytanic acid production is very slow, and most phytol was remained in the rumen as the ester form.

5.3　Factors on the changes of phytanic acid content in milk

Feeding management, diets, species and genetic factors were

important factors for the quality and quantity of milk production (Brito and Broderick, 2007; Wullepit et al., 2009; Sova et al., 2013; Canaza-Cayo et al., 2016). In this study, phytanic acid in milk was higher when the cows were fed the diets with higher phytol content, although the phytanic acid content was lower than those reported in the other studies (Markus et al., 2012). Therefore, it was confirmed that the dietary composition and quality of forages were important factors for determining phytanic acid content in milk. Markus et al. (2012) also reported that phytanic acid content in milk increased when the cows were fed a high herbage based diet. Phytanic acid content in milk would be affected by the herbage applied at different fertilization levels or harvested at different stages.

In Chapter 3, the production ratio of ruminal phytanic acid for silage was found to be 17%~38%. Based on the conversion ratio of dietary phytol to milk phytanic acid (2.6%), about 7%~15% of phytanic acid produced in the rumen would be secreted in milk (2.6 divided by 17%~38%). Phytanic acid was also found in feces, therefore, phytanic acid might exist in blood, body fat, milk, rumen, duodenum and feces. Although it was still unclear the phytanic acid content in each segment, phytanic acid in plasma might be higher than that in milk. In this study, the convertion ratio of phytanic acid in milk was only 2.6%. In Chapter 3, most phytol could not be converted into phytanic acid in the rumen, even though 70% of dietary phytol was estimated to be utilized in the digestive tract. Therefore, studies on the increasing conversion rate of phytol residue into phytanic acid is necessary.

5.4 Conclusion

Fertilization levels and harvesting stages are main factors affecting the contents of photosynthetic pigments and phytol in herbages, and their contents decrease after dried under natural conditions. Although chlorophyll and β-carotene contents in herbages decrease during ensiling, ensiling is an effective method to preserve lutein and phytol.

Dietary phytol content is the main factor for phytanic acid production in the rumen. Although increasing dietary phytol content could increase phytanic acid production, the conversion ratio from dietary phytol to ruminal phytanic acid is low and most phytol is remained in the rumen.

Phytanic acid content in cow's milk is affected by the phytol content of the diets. However the conversion ratio from dietary phytol to milk phytanic acid is not affected by the dietary phytol content, and estimated to be 2.6%.

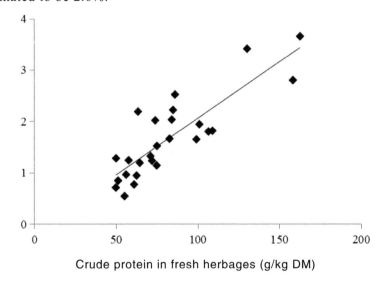

Figure 5.1 Relationship between crude protein content (X) and total phytol content (Y) in fresh herbages. $Y=0.022X-0.15$; $r^2 =0.72$.

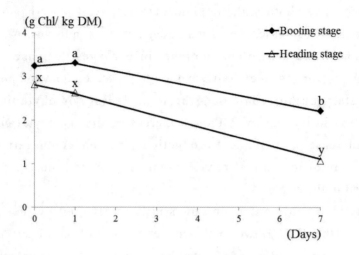

Figure 5.2 Chlorophyll content in Italian ryegrass at booting and heading stage with advancing drying period. (a, b: $P<0.05$; x, y: $P<0.05$).

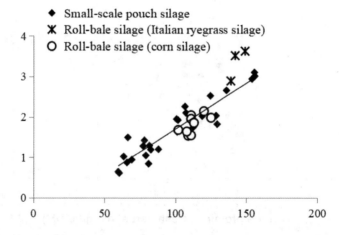

Figure 5.3 Relationship between crude protein content (X) and total phytol content (Y) in silage. $Y=0.077 X-1.95$; $r^2 =0.88$ (small–scale silage).

Chapter 6 General summary

Ruminants have obvious characters to produce milk and meat through utilizing forages. Therefore, it is necessary to explore the potential value of forages and to maximize their utilization of forages. Recently, consumers have shown increased concern for livestock products enriched with bioactive compounds that impact on human health. The content of such functional compounds in ruminant products is considered to relate to their content in diets consumed by animals. Abundant chlorophyll and carotenoids in green forage are considered to be related to the function of ruminant products for human health. The ruminal degradation of chlorophyll in ingested forages liberates phytol moiety which is microbial metabolized to phytanic acid, a natural ligand of peroxisome proliferator-activated receptors. This phytanic acid particularly appears in meat and milk produced by ruminants, and presumably has positive effects on human health.

The study aimed to investigate the factors affecting phytol contents in Italian ryegrass (IR) herbages, and extent of phytanic acid production in ruminants.

In the Chapter 2, changes of photosynthetic pigments and phytol content of herbage were investigated through 4 experiments. In Experiment 1, three rates nitrogen (N) fertilization levels (0 kg N/ha; 60 kg N/ha; and 120 kg N/ha) were applied for IR, and the contents of chemical components and photosynthetic pigments (-carotene, lutein and chlorophylls) in fresh herbage and hay prepared by natural conditions were measured. The crude protein (CP), ether extract (EE), photosynthetic pigments and phytol in IR (fresh herbage and hay) linearly increased with increasing N fertilization levels, and decreased obviously for hay preparation (chlorophylls: 40%~70%, phytol: 25%~47%, -carotene: 72%~90%).

In Experiment 2, the time course changes of carotenoid, chlorophyll and phytol during ensiling were determined for IR herbage. The IR harvested at the heading stage (May 2014) were allowed to wilt under natural conditions for 1 day, and then ensiled using a small-scale pouch system (12 bags). Three bags were destructively analyzed each week to determine the contents of photosynthetic pigments and phytols (esterified and free forms) over a 5-week period (every three bags were unsealed at the 1st, 2nd, 3rd and 5th week, respectively). -carotene content decreased at 2 week after ensiling, while the lutein content did not change significantly. Although the chlorophyll content decreased rapidly in the first week of ensiling, the total phytol content barely changed over the five weeks. During ensiling, the chlorophyll decomposed to pheophytine, then, further degraded to pheophorbide and phytol. Re-esterification of the released phytol might have contributed the stable phytol content during the ensiling of IR.

In Experiment 3, investigate the effects of N fertilizer application and harvesting stage on the contents of chlorophyll, phytol and carotenoids in IR herbage before and after ensiling, and the extent of phytol preservation after ensiling. Three rates of N fertilizer (same as Experiment 1) were applied for IR. The herbage harvested at the booting stage (27 weeks of age, April 25, 2014) or heading stage (29 weeks of age, May 9, 2014) were wilted for 1 day, and then ensiled for 60 days using a small-scale pouch system. The experimental results again verified that photosynthetic pigments content in pre-ensiled herbages increased with the extent of N fertilization levels. In addition, the contents of photosynthetic pigments and phytol at booting stages were higher than those at heading stage. In silage, increasing N fertilizer application also increased the contents of CP, EE, the photosynthetic pigments and their derivatives (pheophytins and pheophorbides), while harvesting stage did not affect the contents

Chapter 6 General summary

of -carotene, chlorophylls and pheophorbides. Lutein and phytol contents were also higher in the silage at the booting stage or grown under higher N fertilizer treatment. In the pre-ensiled herbage, the molar content of phytol was higher than those of the chlorophyll content. A part of the phytol might be derived from the substances other than chlorophyll. N fertilizer application and early harvesting of herbage increased carotenoids and phytol contents in IR silage. Lutein and phytol in IR herbages were indicated to be well preserved during ensiling.

In Experiment 4, investigate the effects of adding lactic acid bacteria (LAB) on the changes in photosynthetic pigments and phytol contents during ensiling of IR grown under different N fertilization levels. The IR herbages grown with three levels of N fertilizer levels (same as Experiment 1) were harvested at the heading stage, and were allowed to wilt under natural conditions for 1 day. The chopped wilted herbages were applied two different treatments: (1) without additive, (2) LAB addition (5 mg/kg fresh grass), then they were ensiled for 60 days using a small-scale pouch system. After ensiling, decreasing silage pH with addition of LAB increased -carotene content, but, unaffected phytol content in silage.

In Chapter 3, the ruminal phytanic acid production from herbage phytol was explored in incubation experiments, with fresh herbage (Experiment 5) and silage (Experiment 6 and 7). The herbage in both experiments were harvested in various N fertilization levels and harvest stages. Two adult wethers fitted with rumen cannulae were used as donors of ruminal fluid. The wethers were fed basal diets of 50% hay and 50% concentrate diets at maintenance energy level. The rumen fluid was collected after 2 hours feeding in the morning and used for incubation. After 48 hours incubation, the phytanic acid production was higher for both herbages at higher N fertilization levels and at booting stage. The ratio of phytanic acid production to total phytol was higher for silage (15%~36%) compared with those for fresh herbages (12%~17%).

In Chapter 4, phytanic acid content in milk was investigated for cows fed total mixed ration (TMR) containing IR silage and corn silage. 17 cows were used in the experiment. Three periods (each period was 21 days) were applied for the experiment. In the period 1 and period 3, the cows were fed corn silage TMR, while the cows were fed IR silage TMR in the period 2. Phytanic acid content in milk was higher for cows fed IR silage TMR compared with those fed corn silage TMR. However, conversion ratio from dietary phytol to milk phytanic acid was estimated to be only 2.6%. There were no differences in milk yield and milk component contents between cows fed IR silage TMR and corn silage TMR.

In summary, the results of the study indicated that higher N fertilization levels or harvest at early stage are an effective way to improve the phytol content in the herbages. Although the phytol content in hay prepared under natural conditions decreased largely, ensiling could effectively preserve phytol in herbage independent on the fermentation quality of silage. The forages containing higher phytol could produce higher phytanic acid in the rumen. However, the phytanic acid production in the rumen is relatively low, and most phytol residue may be remained in the rumen. Milk yield and milk component are not affected by phytol content in diets, however, phytanic acid in milk is higher in cows fed phytol rich diets.

References

[1] Ackman, R. G., Hansen, R. P., 1967. The occurrence of diastereomers of phytanic and pristanic acids and their determination by gas-liquid chromatography. Lipids 2, 357-362.

[2] Alfonso, L., Silke, M., Heina, N., 2001. Phytanic acid and docosahexaenoic acid increase the metabolism of all-transretinoic acid and CYP26 gene expression in intestinal cells. Biochim Biophys Acta. 1521 (1), 97-106.

[3] Allen, N. E., Grace, P. B., Ginn, A., Travis, R. C., Roddam, A. W., Appleby, P. N., Key, T., 2008. Phytanic acid: measurement of plasma concentrations by gas-liquid chromatography-mass spectrometry analysis and associations with diet and other plasma fatty acids. Brit. J. Nutr. 99, 653-659.

[4] AOAC, 1990. Official Methods of Analysis, 15th ed. Association of Official Analytical Chemists, Arlington, VA, USA.

[5] Ballet, N., Robert, J. C., Williams, P. E. V., 2000. Vitamins in forages, in: Givens, D. I., Owen, E., Axford, R. F. E., Omed, H.M. (Eds.), Forage Evaluation in Ruminant Nutrition. CABI Publishing, Wallingford, UK, pp. 399-431.

[6] Beaulieu, R., Seoane, J. R., Savoie, P., Tremblay, D., Tremblay, G. F., Theriault, R., 1993. Effects of dry matter content on the nutritive value of individually wrapped round-bale timothy silage fed to sheep. Can. J. Anim. Sci. 73, 343-354.

[7] Belanche, A., Fuente, G., Pinloche, E., Newbold, C. J., Baleells, J., 2012. Effect of diet and absence of protozoa on the rumen microbial community and on the representativeness of bacterial

fractions used in the determination of microbial protein synthesis. J. Anim. Sci. 90, 3924-3936.

[8] Belury, M. A., 2002. Dietary conjugated linoleic acid in health: physiological effects and mechanisms of action. Annu. Rev. Nutr. 22, 505-531.

[9] Blanch, M., Carro, M. D., Ranilla, M. J., Viso, A., Vazuez-Anon, M., Bach, A., 2016. Influence of a mixture of cinnamaldehyde and garlic oil on rumen fermentation, feeding behavior and performance of lactating dairy cows. J. Dairy Sci. 219, 313-323.

[10] Bourn, D., Prescott, J., 2002. A comparison of the nutritional value, sensory qualities, and food safety of organically and conventionally produced foods. Crit. Rev. Food Sci. Nutr. 42, 1-34.

[11] Bowker, B. C., Grant, A. L., Swartz, D. R., Gerrard D. E., 2004. Myosin heavy chain isoforms influence myofibrillar ATPase activity under simulated postmortem pH, calcium, and temperature conditions. Meat Sci. 67 (1), 139-147.

[12] Breese, E. L., 1983. Exploitation of genetic resource through breeding: Lolium species. in: Mclover, G., Bray, R. A. (Eds). Genetic Resources of Forage Plants CSIRO. Australia, pp. 275-288.

[13] Brito, A. F., Broderick, G. A., 2007. Effects of different protein supplements on milk production and nutrient utilization in lactating dairy cows. J. Dairy Sci. 90, 1816-1827.

[14] Bruhn, J. C., Oliver, J. C., 1978. Effect of storage on tocopherol and carotene concentrations in alfalfa hay. J. Dairy Sci. 61, 980-982.

[15] Buxton, D. R., 1996. Quality-related characteristics of forages as influenced by plant environment and agronomic factors. Anim. Feed Sci. Technol. 59, 37-49.

[16] Cai, Y., Benno, Y., Ogawa, M., Kumai, S., 1999. Effect of applying lactic acid bacteria isolated from forage crops on fermentation characteristics and aerobic deterioration of silage. J. Dairy Sci. 82, 520-526.

[17] Canaza-Cayo, A. W., Cobuic, J. A., Lopes, P. S., Rorres, R.

A., Martins, M. F., Daltro, D. S., Barbosa da Silva, M.V. G., 2016. Genetic trend estimates for milk yield production and fertility traits of the Girolando cattle in Brazil. Livest Sci. 190, 113-122.

[18] Chen, L., Guo, G., Yu, C., Zhang, J., Shimojo, M., Shao, T., 2015. The effects of replacement of whole-plant corn with oat and common vetch on the fermentation quality, chemical composition and aerobic stability of total mixed ration silage in Tibet. Anim. Sci. J. 86, 69-76.

[19] Cherdthong, A., Wanapat, M., Rakwongrit, D., Khota, W., Khantharin, S., Tangmutthapattharakun, G., Kang, S., Foiklang, S., Phesatcha. K., 2014. Supplementation effect with slow-release urea in feed blocks for Thai beef cattle-nitrogen utilization, blood biochemistry and hematology. Trop. Anim. Health Prod. 46, 293-298.

[20] Chew, B. P., Wong, M., Wong, T. S., 1996. Effects of lutein from marigold extract on immunity and growth of mammary tumors in mice. Anticancer Res. 16, 3689-3694.

[21] Choea, J. H., Choia, Y. M., Leea, S. H., Shina, H. G., Ryua, Y. C., Hongb, K. C., Kim, B. C., 2008. The relation between glycogen, lactate content and muscle fiber type composition, and their influence on postmortem glycolytic rate and pork quality. Meat Sci. 80 (2), 355-362.

[22] Choia, Y. M., Ryu, Y. C., Kim, B. C., 2007. Influence of myosin heavy-and light chain isoforms on early postmortem glycolytic rate and pork quality. Meat Sci.76 (2), 281-288.

[23] Comino, L., Tabacco, E., Righi, F., Revello-Chion, A., Quarantelli, A., Borreani, G., 2014. Effects of an inoculant containing a Lactobacillus buchneri that produces ferulate-esterase on fermentation products, aerobic stability, and fibre digestibility of maize silage harvested at different stages of maturity. Anim. Feed Sci. Technol. 198, 94-106.

[24] Crisan, M., Casteilla, L., Lehr, L., Carmona, M., Paoloni-Giacobino, A., Yap, S., Sun, B., Léger, B., Logar, A., Pénicaud, L., Schrauwen, P., Cameron, S. D., Russell, A. P.,

Péault, B., Giacobino, J. P., 2008. A reservoir of brown adipocyte progenitors in human skeletal muscle. Stem Cells, 26（9）, 2425-2433.

[25] Csupor, L., 1971. Das phytol in vergilbten Blättern. Planta. Med. 19, 37-41.

[26] Dahl, G. E., Elsasser, T. H., Capuco, A.V., Erdman, R. A., Peters, R. R., 1997. Effects of a long daily photoperiod on milk yield and circulating concentrations of insulin-like growth factor-I. J. Dairy Sci. 80, 2784-2789.

[27] Dewhurst, R. J., Scollan, N. D., Lee, M. R. F., Ougham, H. J., Humphreys, M. O., 2003. Forage breeding and management to increase the beneficial fatty acid content of ruminant products. Proc. Nutr. Soc. 62, 329-336.

[28] Dierking, R. M., Kallenbach, R. L., Roberts, C. A., 2010. Fatty acid profiles of orchardgrass, tall fescue, perennial ryegrass, and alfalfa. Crop Sci. 50, 391-402.

[29] Dong, C. F., Shen, Y. X., Ding C. L., Xu, N. X., Cheng, Y. H., Gu, H.R., 2013. The feeding quality of rice (Oryza sativa L.) straw at different cutting heights and the related stem morphological traits. Field Crop. Res. 141, 1-8.

[30] Ellinghaus, P., Wolfrum, C., Assmann G., Spener, F., Seedorf, U., 1999. Phytanic acid activates the peroxisome proliferator-activated receptor alpha (PPARalpha) in sterol carrier protein 2-/sterol carrier protein x-deficient mice. J Biol Chem. 274（5）, 2744-2772.

[31] Endo, T., Kubo-nakano, K., Lopez, R. A., Serrano, R. R., Larrinaga, J. A., Yamamoto, S., Honna, T., 2014. Growth characteristics of kochia (Kochia scoparia L.) and alfalfa (Medicago sativa L.) in saline environments. Grassl. Sci. 60, 255-232.

[32] Fraser, M. D., Fychan, R., Jones, R., 2001. The effect of harvest date and inoculation on the yield, fermentation characteristics and feeding value of forage pea and field bean silages. Grass Forage Sci. 56, 218-230.

[33] Garcia, A. D., Olson, W. G., Otterby, D. E., Linn, J. G., Hansen, W. P., 1989. Effects of temperature, moisture, and aeration

on fermentation of alfalfa silage. J. Dairy Sci. 72, 93-103.

[34] Garcia R. P., Antaramian A., Gonzalez, D. L., Villarroya, F., Shimada A., Varela, E. A., Mora, O., 2010. Induction of peroxisomal proliferator-activated receptor γ and peroxisomal proliferator-activated receptor γ coactivator 1 by unsaturated fatty acids, retinoic acid, and carotenoids in preadipocytes obtained from bovine white adipose tissue. J Anim Sci, 88 (5), 1801-1808.

[35] Gloerich, J., van den Brink, D. M., Ruiter, J. P. N., van Vlies, N., Vaz, F. M., Wanders, R. J. A., 2007. Ferdinandusse SMetabolism of phytol to phytanic acid in the mouse, and the role of PPARα in its regulation. J Lipid Res. 48 (1), 77-85.

[36] Gloerich, J., Vlies, N. V., Jansen, G. A., Denis, S., Ruiter, J. P. N., van Werkhoven, M. A., Duran, M., Vaz, F. M., Wanders, R. J. A., Ferdinandusse, S., 2005. A phytol-enriched diet induces changes in fatty acid metabolism in mice both via PPARα-dependent and-inde-pendent pathways. J Lipid Res. 11 (46), 716-726.

[37] Gillespie, C. A., Simpson, D. R., Edgerton, V. R., 1970. High glycogen content of red as opposed to white skeletal muscle fibers of guinea pigs. J Histochem Cytochem. 18 (8), 552-558.

[38] Grimaldi, P. A., 2007. Peroxisome proliferator-activated receptors as sensors of fatty acids and derivatives. Cell Mol. Life Sci. 64, 2459-2464.

[39] Haaland, G. L., Tyrrell, H. F., Moe, P. W., Wheeler, W. E., 1982. Effect of crude protein level and limestone buffer in diets fed at two levels of intake on rumen pH, ammonia-nitrogen, buffering capacity and volatile fatty acid contencration of cattle. J. Anim Sci. 55, 943-950.

[40] Hák, R., Rinderle-Zimmer, U., Lichtenthaler, H. K., Nátr, L., 1993. Chlorophyll a fluorescence signatures of nitrogen deficient barley leaves. Photosynthetica 28, 151-159.

[41] Hashimoto, T., Shimizu, N., Kimura, T., Takahashi, Y., Ide, T., 2006. Polyunsaturated fats attenuate the dietary phytol-induced increase in hepatic fatty acid oxidation in mice. J Nutr, 136

(4), 882-886.

[42] He, J., Watkins, S., Kelley, D. E., 2001. Skeletal muscle lipid content and oxidative enzyme activity in relation to muscle fiber type in type 2 diabetes and obesity. Diabetes, 50 (4), 817-823.

[43] Hellgren, L. I., 2010. Phytanic acid-an overlooked bioactive fatty acid in dairy fat? Ann NY Acad Sci, 1190 (1), 42-49.

[44] Heim, M., Johnson, J., Boess, F., Bendik, I., Weber, P., Hunziker, W., Flühmann, B., 2002. Phytanic acid, a natural peroxisome proliferator-activated receptor (PPAR) agonist, regulates glucose metabolism in rat primary hepatocytes. FASEB J, 16 (7), 718-720.

[45] Heeren, J. A. H., Podesta, S. C., Hatew, B., Klop, G., van Laar, H., Bannink, A., Warner, D., de Jonge, L. H., Dijkstra, J., 2014. Rumen degradation characteristics of ryegrass herbage and ryegrass silage are affected by interactions between stage of maturity and nitrogen fertilisation rate. Anim. Prod. Sci. 54, 1263-1267.

[46] Hellgren, L. I., 2010. Phytanic acid—an overlooked bioactive fatty acid in dairy fat? Ann. N. Y. Acad. Sci. 1190, 42-49.

[47] Hörtensteiner, S., Kräutler, B., 2000. Chlorophyll breakdown in oilseed rape. Photosynth. Res. 64, 137-146.

[48] Hoyt, P. B., 1970. Fate of chlorophyll in soil. Soil Sci. 111, 49-53.

[49] Huyen, N. T., Wanapat, M., Navanukraw, C., 2012. Effect of Mulberry leaf pellet (MUP) supplementation on rumen fermentation and nutrient digestibility in beef cattle fed on rice straw-based diets. Anim. Feed Sci. Technol. 175, 8-15.

[50] Islam, M. R., Garcia, S. C., 2011. Effects of sowing date and nitrogen fertilizer on forage yield, nitrogen- and water-use efficiency and nutritive value of an annual triple-crop complementary forage rotation. Grass Forage Sci. 67, 96-110.

[51] Jensen, S. K., Johannsen, A. K. B., Hermansen, J. E., 1999. Quantitative secretion and maximal secretion capacity of retinol, β-carotene and α-tocopherol into cows' milk. J. Dairy Res. 66, 511-522.

[52] Jeong, J. Y., Kim, G. D., Ha, D. Min., Park, M. J., Park, B. C., Joo, S. T., Lee, C., 2012. YoungRelationships of muscle fiber characteristics to dietary energy density, slaughter weight, and muscle quality traits in finishing pigs. J Anim Sci Tech. 54（3）,175-183.

[53] June, B. P., Mei, G., Gibberd, F. B., Burston, D., Mayne, P. D., McClinchy, J. E., Sidey, M., 1993. Diet and Refsum's disease. The determination of phytanic acid and phytol in certain foods and application of these knowledge to the choice of suitable convenience foods for patients with Refsum's disease. J. Human Nutr. Diet. 6, 295-305.

[54] Johnson, L. M., Harrison, J. H., Davidson D., Mahanna, W. C., Shinners, K., 2003. Corn silage management: effects of hybrid, chop length, and mechanical processing on digestion and energy content.J. Dairy Sci. 86（1）, 208-231.

[55] Kaewlamun, W., Okouyi, M., Humblot, P., Remy, D., Techakumphu, M., Duvaux-ponter, C., Ponter, A. A., 2011. The influence of a supplement of β-carotene given during the dry period to dairy cows on colostrum quality, and β-carotene status metabolites and hormones in newborn calves. Anim. Feed Sci. Technol. 165, 31-37.

[56] Kalač, P., Kyzlink, V., 1979. Losses of beta-carotene in red clover in an acid medium during ensiling. Anim. Feed Sci. Technol. 4, 81-89.

[57] Kalač, P., McDonald, P., 1981. A review of the changes in carotenes during ensiling of forages. J. Sci. Food Agric. 32, 767-772.

[58] Kalač, P., 2012. Carotenoids, ergosterol and tocopherols in fresh and preserved herbage and their transfer to bovine milk fat and adipose tissues: A review. J. Agrobiol. 29, 1-13.

[59] Karlsson, A. H., Klont. R. E., Fernandez, X.,1996. Skeletal muscle fibres as factors for pork quality. Livest Prod Sci. 60（2/3）, 255-269.

[60] Keady, T. W., O'Kiely P., 1996. An evaluation of the effect of rate of nitrogen fertilization of grassland on silage fermentation, in-silo losses, effluent production and aerobic stability. Grass Forage Sci. 51, 350-360.

[61] Kim, G. D., Jeong, J. Y., Jung, E. Y., Yang, H. S., Lim, H. T., Joo, S. T., 2013. The influence of fiber size distribution of type IIB on carcass traits and meat quality in pigs. Meat Sci. 94 (2), 267-273.

[62] King, C., McEniry, J., Richardson, M., O'Kiely, P., 2012. Yield and chemical composition of five common grassland species in response to nitrogen fertiliser application and phenological growth stage. Acta Agric. Scand., Sect. B 62, 644-658.

[63] King, C., McEniry, J., Richardson, M., O'Kiely, P., 2013. Silage fermentation characteristics of grass species grown under two nitrogen fertilizer inputs and harvested at advancing maturity in the spring growth. Grassl. Sci. 59, 30-43.

[64] Koca, N., Karadeniz, F., Burdurlu, H. S., 2006. Effect of pH on chlorophyll degradation and colour loss in blanched green peas. Food Chem. 100, 609-615.

[65] Kopsell D. A., Kopsell, D. E., Curran-Celentano, J., 2007. Carotenoid pigments in kale are influenced by nitrogen concentration and form. J. Sci. Food Agric. 87, 900-907.

[66] Kume, S., Toharmat, T., 2001. Effect of colostral β-carotene and vitamin A on vitamin and health status of new born calves. Livest. Prod. Sci. 68, 61-65.

[67] Larsen, M. K., Frette, X. C., Kristensen, T., Eriksen, J., Seegaard, K., Nielsen, J. H., 2012. Fatty acid, tocopherol and carotenoid content in herbage and milk affected by sward composition and season of grazing. J. Sci. Food Agric. 92, 2891-2898.

[68] Lee, S. H., Joo, S. T., Ryu, Y. C., 2010. Skeletal muscle fiber type and myofibrillar proteins in relation to meat quality. Meat Sci. 86 (1), 166-170.

[69] Lefsrud, M. G., Sorochan, J. C., Kopsell, D. A., Scott Mcelroy, J., 2010. Pigment concentrations among heat-tolerant turfgrasses. Hort. Sci. 45, 650-653.

[70] Leiber, F., Kreuzer, M., Nigg, D., Wettstein, H. R., Scheeder, M. R. L., 2005. A study on the causes for the elevated n-3 fatty acids in cows' milk of alpine origin. Lipids. 40, 191-202.

[71] Lemotte P.K., Keidel, S., Apfel, C. M.,1996. Phytanic acid is a retinoid X receptor ligand. Eur J Biochem. 236（1）, 328-333.

[72] Lichtenthaler, H. K., 1987. Chlorophyll and carotenoids: pigments of photosynthetic biomembranes. Meth. Enzymol. 148, 350-382.

[73] Liljenberg, C., Odham, G., 1969. Gas chromatographic determination of phytol in plant material. Physiol. Plant. 22, 686-693.

[74] Lindqvist, H., Nadeau, E., Jensen, S. K., 2012. Alpha-tocopherol and β-carotene in legume-grass mixtures as influenced by wilting, ensiling and type of silage additive. Grass Forage Sci. 67, 119-128.

[75] Lin, X. J., Shu, G., Zhu, X. T., 2018. Effects of phytol on metabolic enzyme activity of skeletal muscle and muscle fiber type of mice. Chin. J. Anim. Nutr. 30（10）, 4237-4243.

[76] Lin, X. J., Zhu, X. T., Jiang Q. Y., Shu G., 2012. Modulation effects of phytol on adipocytes differentiation and glucolipid metabolism. Chin. J. Anim. Nutr. 24（10）, 1866-1870.

[77] Liu, Q., 2011. Study on the structure and function of glucokinase, elaiophyln glycosyltransferase and leucine-rich repeat kinase 2, PhD thesis. Beijing: Chinese Academy of Sciences, 20-23.

[78] Lv, R. L., EL-Sabagh, M., Obitsu, T., Sugino, T., Kurokawa, Y., Kawamura, K., 2017. Effects of nitrogen fertilizer and harvesting stage on photosynthetic pigments and phytol contents of Italian ryegrass silage. Anim. Sci. J. 88,1513-1522.

[79] Mackie, J. T., Atshaves B. P., Payne, H. R., McIntosh, A. L., Schroeder, F., Kier, A. B., 2009. Phytol-induced hepatotoxicity in mice.Toxicol Pathol. 37（2）, 201-208.

[80] Makoni, N. F., Shelford, J. A., 1993. Changes in lipids, chlorophyll pigments, hot water Insoluble N and pH of Alfalfa during ensiling. J. Sci. Food Agric. 63, 273-280.

[81] Mangels, A. R., Holden, J. M., Beecher, G. R., Forman, M. R., Lanza, E., 1993. Carotenoid content of fruits and vegetables: an evaluation of analytic data. J. Am. Diet. Assoc. 93, 284-296.

[82] Martin, K. R., Failla, M. L., Smith Jr, J. C., 1996.

-Carotene and lutein protect HepG2 human liver cells against oxidant-induced damage. J. Nutr. 126, 2098-2106.

[83] Martin, B., Cornu, A., Kondjoyan, N., Ferlay, A., Verdier-Metz, I., Pradel, P., Rock, E., Chilliard, Y., Coulon, J. B., Berdague, L., 2005. Milk indicators for recognizing the types of forages eaten by dairy cows, in: Hocquette, J. F., Gigli, S. (Eds.), Indicators of Milk and Beef Quality. EAAP Publication no. 112. Wageningen Academic Publishers, Wageningen, The Netherlands, pp. 127-136.

[84] Masoero, F., Gallo, A., Zanfi, C., Giuberti, G., Spanghero, M., 2011. Effect of nitrogen fertilization on chemical composition and rumen fermentation of different parts of plants of three cornhybrids. Anim. Feed Sci. Technol. 164, 207-216.

[85] Masters-Thomas. A., Bailes, J., Billimoria, J. D., Clemens, M. E., Gibberd, F. B., 1980. Heredopathia atactica polyneuritiformis (Refsm's disease): 2. Estimation of phytanic acid in foods. J. Human Nutr. 34, 251-254.

[86] Matile, P., Hörtensteiner, S., Thomas, H., Kräutler, B., 1996. Chlorophyll breakdown in senescent leaves. Plant Physiol. 112, 1403-1409.

[87] Markus, S., Farideh, Y., Walter, V. 2011. Investigating the day-to-day variations of potential marker fatty acids for organic milk in milk from conventionally and organically raised cows. Eur. Food Res. Technol. 232, 167-174.

[88] Markus, S., Nina, L. L., Ernest, C. T., Ensieh, H., Farideh, Y., Walter, V. 2012. Phytanic acid concentrations and diastereomer ratios in milk fat during changes in the cows feed from concentrate to hay and back. Eur. Food Res. Technol. 234, 955-962.

[89] Marta, G., Ma, À. O., Marina, G., Alejandro D, Andrzej. A. S., Anne, L., Domingo, C., 2003. The relationship between pig genetics, myosin heavy chain I, biochemical traits and quality of M.longissimus thoracis. Meat Sci. 65 (3), 1063-1070.

[90] McDougal, E. I., 1948. Studied on ruminant saliva. 1. The

composition and output of sheep's saliva. Biochem. J. 43, 99-109.

[91] McMarty, M. F., 2001. The chlorophyll metabolite phytanic acid is a natural rexinoid -potential for treatment and prevention of diabetes. Med. Hypotheses 56, 217-219.

[92] Michal, J. J., Heirman, L. R., Wong, T. S., Chew, B. P., Frigg, M., Volker, L., 1994. Modulatory effects of dietary -carotene on blood mammary leukocytes function in periparturient dairy cows. J. Dairy Sci. 77, 1408-1421.

[93] Moya, D., Calsamiglia, S., Ferret, A., Blanch, M., Fandino, J. I., Castillejos, L., Yoon, I., 2009. Effects of dietary changes and yeast culture (Saccharomyces cerevisiae) on rumen microbial fermentation of Holstein heifers. J. Anim. Sci. 87, 2874-2881.

[94] Neylon, J. M., Kung, L., 2003. Effects of cutting height and maturity on the nutritive value of corn silage for lactating cows.J Dairy Sci.86 (6), 2163-2169.

[95] Nishino, N., Hiroaki, H., Sakaguchi, E., 2003. Evaluation of fermentation and aerobic stability of wet brewers' grains ensiled alone or in combination with various feeds as a total mixed ration. J. Sci. Food Agric. 83, 557-563.

[96] Nishino, N., Uchida, S., 1999. Laboratory evaluation of previously fermented juice as a fermentation stimulant for lucerne silage.J. Sci. Food Agr. 79, 1285-1288.

[97] Nozière, P., Graulet, B., Lucas, A., Martin, B., Grolier, P., Doreau, M., 2006. Carotenoids for ruminants: from forages to dairy products. Anim. Feed Sci. Technol. 131, 418-450.

[98] NRC, Nutrient requirements of beef cattle, sixth revised ed., National Research Council. National Academy Press, Washington, DC, USA, 2001.

[99] Nurfeta, A., Eik L. O., Tolera, A., Sundstøl, F., 2008. Chemical composition and in sacco dry matter degradability of different morphological fractions of 10 enset (Ensete ventricosum) Varieties. Anim. Feed Sci. Tech.146 (1/2), 55-73.

[100] O-Mara, F. P., Fitzgerald, J. J., Murphy, J. J., Rath, M., 1998. The effect on milk production of replacing grass silage with maize silage in the diet of dairy cows. Livest. Sci. 55, 79-87.

[101] Olson, J. A., 1989. Biological actions of carotenoids. J. Nutr. 119, 94-95.

[102] Ouellet, D. R., Chiquette, J., 2016. Effect of dietary metabolizable protein level and live yeastson ruminal fermentation and nitrogen utilization in lactatingdairy cows on a high red clover silage diet. Anim. Feed Sci. Technol. 220, 73-82.

[103] Paiva, P. G., Del Valle, T. A., Jesus, E. F., Bettero, V. P., Almeida, G. F., Bueno, I. C. S., Bradford, B. J., Renno, F. P., 2016. Effects of crude glycerin on milk composition, nutrient digestibility and ruminal fermentation of dairy cows fed corn silage-based diets. Anim. Feed Sci. Technol. 212, 136-142.

[104] Palmquist, D. L., Lock, A. L., Shingfield, K. J., Bauman, D. E., 2005. Biosynthesis of conjugated linoleic acid in ruminants and humans. Adv. Food Nutr. Res. 50, 179-217.

[105] Park, Y. W., Anderson, M. J., Walters, J. L., Mahoney, A. W., 1983. Effects of processing methods and agronomic variables on carotene contents in forages and predicting carotene in alfalfa hay with near-infrared-reflectance spectroscopy. J. Dairy Sci. 66, 235-245.

[106] Peisker, C., Düggelin, T., Rentsch, D., Matile, P., 1989. Phytol and the breakdown of chlorophyll in senescent leaves. J. Plant Physiol. 135, 428-432.

[107] Peter, J. B., Sawaki, S., Barnard, R. J., Edgerton, V. R., Gillespie, C. A., 1971. Lactate dehydrogenase isoenzymes: distribution in fast-twitch red, fast-twitch white, and slow-twitch intermediate fibers of guinea pig skeletal muscle. Arch Biochem Biophys. 144 (1), 304-307.

[108] Peter, J. B, Barnard, R. J., Edgerton, V. R., Gillespie, C. A., Stempel, K. E.,1972. Metabolic profiles of three fiber types of skeletal muscle in guinea pigs and rabbits. Biochem. 11 (14), 2627-2633.

[109] Porra, R. J., Thompson, W. A., Kriedemann, P. E., 1989.

Determination of accurate extinction coefficients and simultaneous equations for assaying chlorophylls a and b extracted with four different solvents: verification of the concentration of chlorophyll standards by atomic absorption spectroscopy. Biochim. Biophys. Acta 975, 384-394.

[110] Radostits, O. M., Gay, C. C., Blood, D. C., Hinchliffe, K. W., 2007. Veterinary Medicine. A text book of the diseases of cattle, sheep, goats and horses, 10th ed. Saunders, W.B. Ltd., London, UK.

[111] Qi, L. F., Xu, Z. R., 2003. PPAR and its regulation on lipid metabolism.Chin J Vet Med. 37（7）,33-35

[112] Razzaghi, A., Valizadeh, R., Naserian, A. A., Danesh Mesgaran, M., Rashidi, L., 2015. Effects of sucrose and sunflower oil addition to diet of Saanen dairy goats on performance and milk fatty acid profile. Livest. Sci. 173, 14-23.

[113] Reist, M., Erdin, D. K., von Euw, D., Tschuemperlin, K. M., Leuenberger, H., Delavaud, C., Chilliard, H. M., Hammon, H. M., Kuenzi, N., Blum, J. W., 2003. Concentrate feeding strategy in lactating dairy cows: Metabolic and endocrine changes with emphasis on leptin. J. Dairy Sci. 86, 1690-1706.

[114] Reynoso, C. R., Mora, O., Nieves, V., Shimada, A., González de Mejía, E., 2004. β-Carotene and lutein in forage and bovine adipose tissue in two tropical regions of Mexico. Anim. Feed Sci. Technol. 113, 183-190.

[115] Rezaei, J., Rouzbehan, Y., Zahedifar, M., Fazaeli, H., 2015. Effects of dietary substitution of maize silage by amaranth silage on feed intake, digestibility, microbial nitrogen, blood parameters, milk production and nitrogen retention in lactating Holstein cows. Anim. Feed Sci. Technol. 202, 32-41.

[116] Ryu, Y. C., Choi, Y. M., Lee, S. H., Shin, H. G., Choe, J. H., Kim, J. M., Hong, K. C., Kim, B. C., 2008. Comparing the histochemical characteristics and meat quality traits of different pig breeds. Meat Sci. 80（2）,363-369.

[117] Ryu, Y. C., Kim B. C., 2005.The relationship between

muscle fiber characteristics, postmortem metabolic rate, and meat quality of pig longissimus dorsi muscle.Meat Sci. 71（2）, 351-357.

[118] SAS Institute, 2004. SAS/STAT User's Guide, Version 9.1 Edition. SAS Institute, Cary, NC, USA.

[119] Saltin, B., Henriksson, J., Nygaard, E., Andersen, P., Jansson, E.,1977. Fiber types and metabolic potentials of skeletal muscles in sedentary man and endurance runners.Ann NY Acad Sci. 301（1）, 3-29.

[120] Sarmadi, B., Rouzbehan, Y., Rezaei, J., 2016. Influences of growth stage and nitrogen fertilizer on chemical composition, phenolics, in situ degradability and in vitro ruminal variables in amaranth forage. Anim. Feed Sci. Technol. 215, 73-84.

[121] Sazili, A. Q., Parr, T., Sensky, P. L., Jones, S. W., Bardsley, R. G., Buttery, P. J., 2005. The relationship between slow and fast myosin heavy chaincontent, calpastatin and meat tenderness in different o-vine skeletal muscles. Meat Sci. 69（1）,17-25.

[122] Schluter, A., Giralt, M., Iglesias, R., Villarroya, F., 2002. Phytanic acid, but not pristanic acid, mediates the positive effects of phytol derivatives on brown adipocyte differentiation. FEBS Lett. 517（1）, 83-86.

[123] Schlüter, A., Yubero, P., Iglesias, R., Giralt, M., Villarroya, F., 2002. The chlorophyll-derived metabolite phytanic acid induces white adipocyte differentiation. J IASO. 26（9）, 1277-1280.

[124] Seedorf, U., Raabe, M., Ellinghaus, P., Kannenberg, F., Fobker, M., Engel, T., Denis, S., Wouters, F., Wirtz, K.W., Wanders, R. J., Maeda, N., Assmann, G.,1998. Defective peroxisomal catabolism of branched fatty acyl coenzyme A in mice lacking the sterol carrier protein-2/sterol carrier protein-x gene function. Gene Dev.12（8）,1189-1201.

[125] Seideman, S. C., Crouse, J. D., Cross, H. R., 1986. The effect of sex condition and growth implants on bovine muscle fiber characteristics. Meat Sci. 17（2）, 79-95.

[126] Serra, X., Gil, F., Perez, E. M., Oliver, M. A., Vazquez,

J. M., Gisper, M., Diaz, I., Moreno, F., Latorre, R., Noguera, J. L., 1998. A comparison of carcass, meat quality and histochemical characteristics of Iberian (Guadyerbas line) and Landrace pigs. Livest Prod Sci.56 (3), 215-223.

[127] Shao, T., Ohba, N., Shimojo, M., Masuda, Y., 2002. Dynamics of early fermentation of Italian ryegrass (Lolium multiflorum Lam) silage. Asian-Aust. J. Anim. Sci. 15, 1606-1610.

[128] Sieck, G. C., Zhan, W. Z., Prakash, Y. S., Daood, M. J., Watchko, J., 1995. FSDH and actomyosin ATPase activities of different fiber types in rat diaphragm muscle. J Appl Phys. 79 (5), 1629-1639.

[129] Sova, A. D., Leblanc, S. J., Mcbride, B. W., Devries, T. J., 2013. Associations between herd-level feeding management practices, feed sorting, and milk production in freestall dairy farms. J. Dairy Sci. 96, 4759-4770.

[130] Spanghero, M., Zanfi, C., Signor, M., Davanzo, D., Volpe, V., Venerus, S., 2015. Effect of plant vegetative stage and on field drying time on chemical composition and in vitro ruminal degradation of forage soybean silage. Anim. Feed Sci. Technol. 200, 102-106.

[131] Steinberg, D., Herndon, J. H., Uhlendorf, B. W., Mize, C. E., Avigan, J., Milne, G. W. A., 1967. Refsum's disease: nature of the enzyme defect. Science 156, 1740-1742.

[132] Sukhija, P. S., Paimquist, D. L., 1988. Rapid method for determination of total fatty acid content and composition of feedstuffs and feces. J. Agric. Food Chem. 36, 1202-1206.

[133] Kitareewan, S., Burka, L. T., Tomer, K. B., Parker, C., E., Deterding, L. J., Stevens, R. D., Forman, B. M., Mais, D. E., Heyman, R. A., McMorris, T., Weinberger, C., 1996. Phytol metabolites are circulating dietary factors that activate the nuclear receptor RXR. Mol Biol Cell. 7 (8), 1153-1166.

[134] Suzuki, T., Kamiya, Y., Tanka, M., Hattori, I., Sakaigaichi, T., Terauchi, T., Nonaka, I., Terada, F., 2014. Effect of fiber content of roughage on energy cost of eating and rumination in

Holstein cows. Anim. Feed Sci. Technol. 196, 42-48.

[135] Takahashi, N., Kawada, T., Goto, T., Yamamoto, T., Taimatsu, A., Matsui, N., Kimura, K., Saito., M., Hosokawa, M., Miyashita, K., Fushiki, T., 2002. Dual action of isoprenols from herbal medicines on both PPAR [gamma] and PPAR [alpha] in 3T3-L1 adipo-cytes and HepG2 hepatocytes. FEBS Lett. 514 (2), 315-322.

[136] Takeda, Y., Saito, Y., Uchiyama, M., 1983. Determination of pheophorbide a, pyropheophorbide a and phytol. J. Chromatogr. A 280, 188-193.

[137] Tian, J. P., Yu, Y. D., Yu, Z., Shao, T., Na, R., Zhao, M. M., 2014. Effects of lactic acid bacteria inoculants and cellulose on fermentation quality and in vitro digestibility of Leymus chinensis silage. Grassl. Sci. 60, 199-205.

[138] Tsuyoshi, G., Nobuyuki, T., Sota, K., Kahori, E., Shogo, E. Tatsuya, M., Tohru, F., Teruo, K., 2005. Phytol directly activates peroxisome proliferator-activated receptor [alpha] (PPAR [alpha]) and regulates gene expression involved in lipid metabolism in PPAR [alpha-expressing HepG2 hepatocytes. Biochem Bioph Res Co. 337 (2), 440-445.

[139] Uchida, S., Kim, K. H., Yun I. S., 1989. The effect of wilting on silage making from the viewpoint in connection with monsoon Asia. Asian-Austral. J. Anim. Sci. 2, 43-49.

[140] Ueda, Y., Asakuma, S., Miyaji, M., Akiyama, F., 2016. Effect of time at pasture and herbage intake on profile of volatile organic compounds of dairy cow milk. Anim. Sci. J. 87, 117-125.

[141] Van Ranst, G., Fievez, V., Vandewalle, M., De Riek, J., Van Bockstaele, E., 2009. Influence of herbage species, cultivar and cutting date on fatty acid composition of herbage and lipid metabolism during ensiling. Grass Forage Sci. 64, 196-207.

[142] Van Soest, P. J., Robertson, J. B., Lewis, B. A., 1991. Methods for dietary fiber, neutral detergent fiber, and non-starch polysaccharides in relation to animal nutrition. J. Dairy Sci. 74, 3583-3597.

[143] Vetter, W., Schroder, M., 2010. Concentrations of phytanic

acid and pristanic acid are higher in organic than in conventional dairy products from the German market. Food Chem. 119, 746-752.

[144] Verhoeven, N. M., Wanders, R. J. A., Poll-The, B. T., Saudubray, J. M., Jakobs, C., 1998. The metabolism of phytanic acid and pristanic acid in man: a review. J. Inher. Metab. Dis. 21, 697-728.

[145] Vezitskii, A., 2000. Chlorophyll a formation in etiolated rye seedlings as dependent on the concentration of infiltrated chlorophyllide. Russ. J. Plant Physiol. 47, 499-503.

[146] Wang, J. J., Peng J. L., Zhu, X. T., Wang, L. N., Wang, S. B., Gao, P., Shu, G., Jiang Q. Y., 2014. Effects of phytol on skeletal muscle fiber type and meat quality of fattening pigs. Anim. Husb. Vet. Med. 46, 134-136.

[147] Wang, L., 2016. The effect of muscle fiber types and their differences in matabolic enzymes activity on postmortem tenderness of yak meat. Chinese Journal of Animal Nutrition, volume 30. Master's thesis, Lanzhou: Gansu Agricultural University, 37-38.

[148] Wanders, R. J., Komen, J., Ferdinandusse, S., 2011. Phytanic acid metabolism in health and disease. Biochim. Biophys. Acta. 1811, 498-507.

[149] West, J. W., Hill, G. M., Utley, P. R., 1993. Peanut skins as a feed ingredient for lactating dairy cows. J. Dairy Sci. 76, 590-599.

[150] Whitehead, D. C., 1995. Grassland Nitrogen. CAB International, Wallingford, UK.

[151] Wilkinson, J. M., 2005. Silage. Chalcombe. Publications. Lincoln. 1-254.

[152] Woolford, M. K., 1990. The detrimental effects of air on silage. J. Appl. Bacteriol. 68, 101-116

[153] Wullepit, N., Raes, K., Beerda, B., Veerkamp, R. F., Fremaut, D., De Smet, S., 2009. Influence of management and genetic merit for milk yield on the oxidative status of plasma in heifers. Livest. Sci. 123, 276-282.

[154] Xu, C. Z., Wang, H. F., Yang, J. Y., Wang, J. H., Duan, Y., Wang, C., Liu, J. X., Lao, Y., 2014. Effects of feeding lutein

on production performance, antioxidative status, and milk quality of high-yielding dairy cows. J. Dairy Sci. 97, 7144-7150.

[155] Zapata, M., Rodríguez, F., Garrido, J. L., 2000. Separation of chlorophylls and carotenoids from marine phytoplankton: a new HPLC method using a reversed phase C_8 column and pyridine-containing mobile phases.Mar. Ecol. Prog. Ser. 195, 29-45.

[156] Zhang, K. C., Ji, H. H., 2007. Process on research and development of extraction and activity determination of succinate dehydrogenase.China medicine, 5(10), 32-33.

[157] Zhang, L. X., Cooney, R. V., Bertram. J. S., 1991. Carotenoids enhance gap junctional communication and inhibit lipid peroxidation in C3H/10T1/2 cells: Relationship to their cancer chemopreventive action. Carcinogenesis. 12, 2109-2114.

[158] Zhao, D., Raja Reddy, K., Kakani, V. G., Read, J. J., Carter, G. A., 2003. Corn(Zea mays L.) growth, leaf pigment concentration, photosynthesis and leaf hyperspectral reflectance properties as affected by nitrogen supply. Plant Soil. 257, 205-218.

[159] Zomer, A. W. M., Van der Burg, B., Jansen, G. A., Wanders, R. J. A., Poll-The, B. T., Van der Saag, P. T., 2000. Pristanic acid and phytanic acid: naturally occurring ligands for the nuclear receptor peroxisome proliferator-activated receptoJ. Lipid Res 41, 1801-1807.

[160] Zhu, Y., Nishino, N., Kishida, Y., Uchida, S., 1999. Ensiling characteristics and ruminal degradation of Italian ryegrass and Lucerne silages treated with cell wall-degrading enzymes. J. Sci. Food Agric. 79, 1987-1992.

Abbreviations

AMS: automatic milking systems
A/P: the ratio of acetic acid to propionic acid
BUN: blood urea nitrogen
BW: body weight
CP: crude protein
CS: corn silage
d: day
DIM: days in milk
DM: dry matter
DMI: dry matter intake
DP: disappeared phytol
EE: ether extract
FA: fatty acid
FM: fresh matter
h: hour
H: harvesting stage
IR: Italian ryegrass
LA: lactic acid
LAB: lactic acid bacteria
min: minute
N: nitrogen fertilization
ND: not detected
NDFom: neutral detergent fiber exclusive of residual ash
NEFA: non-esterified fatty acids
NFC: non-fibrous carbohydrate

N*H: interaction effect between N and H
N*T: interaction effect between N and T
PA: phytanic acid
RP: disappeared phytol
SEM: standard error of the mean
SNF: solids not fat
T: additives treatments
T-CHO: total-cholesterol
TDN: total digestible nutrients
TG: triglyceride
TMR: total mixed rations
UV: ultraviolet rays
VFA: volatile fatty acid;
Y: year

ACKNOWLEDGMENTS

This work was supported by the Japan Society for the Promotion of Science (JSPS) KAKENHI (No. 26292138) and the Hainan Provincial Natural Science Foundation of China (No. 322RC774)